Geographies of Nature

For Caroline, Dylan and Nate, for teaching
me a looser kind of sense

Geographies of Nature

societies, environments, ecologies

Steve Hinchliffe

SAGE Publications
Los Angeles · London · New Delhi · Singapore

First published 2007

 SAGE Publications Ltd
1 Oliver's Yard
55 City Road
London EC1Y 1SP

SAGE Publications Inc.
2455 Teller Road
Thousand Oaks, California 91320

SAGE Publications India Pvt Ltd
B 1/I 1 Mohan Cooperative Industrial Area
Mathura Road
New Delhi 110 044

SAGE Publications Asia-Pacific Pte Ltd
33 Pekin Street #02-01
Far East Square
Singapore 048763

Library of Congress Control Number: 2007925375

British Library Cataloguing in Publication data

A catalogue record for this book is available from the
British Library

ISBN 978-1-4129-1048-4
ISBN 978-1-4129-1049-1 (pbk)

Typeset by C&M Digital (P) Ltd., Chennai, India
Printed in India at Replika Press Pvt. Ltd.
Printed on paper from sustainable resources

Contents

Acknowledgements

This book is the product of many places, people, friends, colleagues, books, articles, fields, rivers, plants, conversations, arguments, animals – some of these are quoted in the book, some are described, many will recognize elements of themselves without necessarily being granted the formal recognition they deserve. But my thanks to all those who have inspired me over the years and taken the time to engage with my misreading and misunderstandings of their words and deeds in order to set me right. That I am not right is of course partly inevitable, but something for which I take full responsibility.

The theme of this book, geographies of, or spaces for, nature is indicative of the practice of borrowing and translating that goes on in academic worlds – the term 'Spaces for Nature' had a former life as the name for a project which aimed to increase the number of local nature reserves that existed in urban Birmingham. Being involved in the early stages of that project and in thinking what it might mean for nature to have spaces, inspired some of the arguments that are worked out in this book.

Some of the chapters emerged from projects with others, notably with people involved in the Habitable Cities Project at the Open University (Sarah Whatmore, Matthew Kearnes, Monica Degen and staff at the Birmingham and Black Country Wildlife Trust and CSV Environment). Meanwhile ongoing work with Nick Bingham on biosecurities is included in Chapter 7. Thanks to all these people for helping to generate these and other materials. Work on habitable cities was kindly funded by the UK's Economic and Social Research Council (No. R00239283). I have also benefited greatly from small grants from the Open University, enabling me to carry out fieldwork, and from study leave time made available by the Geography Discipline and my colleagues at the OU.

Special thanks also go to the following people who have read or commented on large or small parts of this book in earlier or later drafts – they include Nick Bingham, Lucila Lahitou, Nigel Clark, John Law, Ingunn Moser, Kristin Asdal and Annemarie Mol.

In addition, there are many people who in recent years have shaped my thinking. They include John Allen, Michael Pryke, Doreen Massey, Sarah Whatmore, David Featherstone, David Papadopoulos, Clive Barnett, Kevin Hetherington, Steve Pile, Nigel Thrift, David Demeritt, Noel Castree, Eric Laurier, Gail Davis, Jacquie Burgess, Andrew Blowers, Carolyn Harrison, Tony Phillips, Glyn Williams, Martin Parker, Jennifer Wolch, Chris Wilbert,

Chris Philo, Jenny Price, Beth Greenhough, Emma Roe, Gareth Enticott, Jeanette Pols, Martin Gren, Mikael Jonasson, Ian Cook, Michael Crang, Paul Harrison, Brendan Gleeson, Sue Owens, Richard Cowell, Adrian Passmore, James Evans, Bronislaw Szersynski, Phil MacNaghten, Claire Waterton, Brian Wynne, Parvati Raghuram, George Revill, and Alistair Phillips.

At Sage I would like to thank Robert Rojek for running with the initial idea and for his encouragement and patience, also Katherine Haw and Susan Dunsmore for their hard work at the editing stage.

Finally, and most importantly, Caroline, Dylan and Nate have taught me a great deal, enchanted me at every turn and have made writing a joyful possibility.

List of Figures

Introduction

From there to here
From here to there
Funny things are everywhere.
(Dr Seuss, *Red Fish, Blue Fish*)

Simply put, this book is about how nature is 'done', how it is practised, how it materializes as an active partner in and through those practices. Perhaps, unlike many other volumes, I am not especially concerned here with how nature is imagined, represented, thought or conceived. Rather, imagining, representing and thinking are treated as activities which take their place alongside many other practices (like growing, infecting, digging, counting), some of which do not have people at their centre. This last point is crucial. For there are many other accounts of nature as produced and practised, within landscape studies, sociology, psychology, political economy and human geography, for example, which tend to reduce nature to not much more than a malleable mass to be shaped at will or at the behest of cultural, economic or political forces and contestations. *Geographies of Nature* is more even-handed, arguing that non-humans are lively and dynamic colleagues in the making of worlds. Yet in being even-handed I do not mean to suggest that the book is politically neutral. Indeed, the politics of nature become more pressing once the contributions to those politics become 'more than human' (the phrase comes from Whatmore, 2004).

So why 'geographies' of nature? People have long been used to the idea of natural history. Landscapes and species are often given a history, although usually one that emphasizes their interrelations with cultures and peoples. What, then, about a natural geography? One way to do this would be to rehearse a rather tired geography which talks about nature's gradual or sudden retreat to the peripheries of modern societies. Another approach would be to develop geographies where natures and societies are interwoven in a variety of different ways with a variety of different effects. This would be to note how natures vary from place to place, that there are cultures of nature (for exemplars, see Jardine et al., 1996; Livingstone, 2003; Matless, 1998; Wilson, 1992). But even here, nature can remain rather passive and ironically a-spatial. Indeed, it usually turns out that it is culture that varies through history and from place to place, while nature stays much the same. So another possibility would be to develop geographies of natures wherein natures are made up of many different practices, all of which are implicated in the continual shaping of those natures. A woodland, for example, will be practised by and with many different species, people, habits, artefacts, in many different places (from soil horizons to government offices, from prevailing winds to balance sheets). So a natural history of the woodland could be written, but so too could a natural geography. In place of a contested history of nature, we can, to put the point too baldly, give nature a complex present (Mol, 2002: 43), or better, start to map together or diagram the histories and geographies (the space-times, Massey, 2005) of natures. The result is, I hope, a sensitivity to the multiplicity that is the very stuff of the world. It's an ethos, or mode of attention (as Donna Haraway (2003) puts it), which allows for more of the world to colour and affect the way that world is made and remade.

So this is a positive, affirmative account of nature, of spaces for nature. And yet, rumours of the death of nature have been around for a long time. According to some, such as the critical feminist historian Caroline Merchant (1990), the death knell was sounded some time ago, at the birth of the modern world and in particular with the advent of 'Western' seventeenth-century science and technology. More famously, in terms of popular literature, the end of nature has been reported by Bill McKibben (2003). For him the expansion of human influence to every corner of the globe, and in particular the changes wrought on such mammoth systems as climate and oceans, has meant that if we went looking for nature, we would be hard pressed to find anything that was untouched by human hands. More recently, the sociologist Bruno Latour (2004b) has attempted to nail the coffin firmly shut by suggesting, even insisting, that we abolish Nature. For Latour, Nature does no more and no less than act as a convenient foil. It allows so-called modern societies to act as if the nonhuman world were mute and malleable material. This has two silencing effects:

1 Animals, plants, human bodies, tectonic plates, stem cells, proteins – all, according to this modern version of the world (Latour, 1993), do what they are told. They behave. They perform to pre-written scripts, and obey the rules and laws of the game. The only thing left in this rather disenchanted world (Bennett, 2001) that is remarkable, that is worthy of remark or literally is able to re-mark or re-script itself, is human ingenuity. And this ingenuity allows people to read all the scripts that these poor others are forced to simply act out. Once read, humans can tell you what is going to happen next (for everything performs to (the) type), or even get in there and change bits of the script (think of genetic modification), so that changes can occur in determinable and predictable ways.

2 As we have already heard, natural objects are reliable followers of scripts and thereby their limited behaviour forms the basis for determinate laws of nature. Such laws are generally non-negotiable, they just are. Which

means that human beings can do one of two things. Either they can say, 'All this stuff that just is, that simply behaves, is not important to the world of politics, to the world of making complex decisions about how to live together. We can therefore leave all these embryos, rivers, machines, molecules of carbon dioxide out of politics (and out of human geography).' Or they can say, 'If we leave things to people they will argue incessantly about how to go on, so let us look for natural laws, non-negotiable truths, that can shape our actions. Let us bring the bickering to a halt by saying this is the way things are, this is natural, and then find ways of living accordingly.'

None of this is satisfactory. As I argue throughout the book, performing to a script happens, but not as often as this version of mute nature suggests. And politics, people, animals, rivers, tectonic plates – none of them are well served by imagining that all power to decide and judge resides on one side of a human–nonhuman divide (or even that a politics is best implemented by skilfully moving from one side to another in the heat of action, hoping that nobody will notice).

Geographies of Nature is written out of a great deal of sympathy for Latour's work. And yet, unlike this work, there is less of a compulsion to finish nature off in these pages. As I have already indicated this is a book arguing *for* nature. In other words, there's a belief that the word can still do some work. (In the text I sometimes use 'Nature' with a capital N when reference is made to the idea of a fixed and single world, totally outside systems of understanding and acting. I prefer to use 'nature', small n, to denote that natures are made but not in ways that are reducible to human meaning systems.) In the following pages, nature (certainly demoted from the capital Nature) is alive and well and living in inner-city Birmingham, in subtropical Africa, in laboratories, on farms, in the offices of European governments, on allotments, and so on. In this I echo and expand on Thrift's (2000) sense of nature as biopolitical domain – to wit, that far from being dead and buried, nature is currently being practised anew (see also Franklin, 2002). But, given that nature is not what we have imagined it to be, fixed in its identity and unrelated to societies, a crucial question remains as to what kinds of spaces there are for nature. What sorts of spaces can overcome the tendency to either assume nature is dead, or assume that it exists, neatly bounded, incarcerated in a self-sealed cell? How can we productively find proximities and distances from, in, for, to nature, in order to avoid swamping all and every nonhuman and human being with cultural artifice, prevent over-sentimentalizing others and/or reduce everyone and everything to a relational force field wherein most of them matter little? The following chapters will explore various natural relations, topologies of nature, nature and 'difference' in order to tease out a multiplicity of spaces for nature.

So the book aims to re-figure what nature is and can be and at the same time experiment with the sorts of spaces that we can generate *for* those natures (human and nonhuman). Throughout it will be argued that this experiment matters. For without this unsettling and re-settling of nature's spaces we will continue to produce unjust politics. A politics that consigns people, animals, plants, and the various assemblies that are made up of more or less of these, to unsatisfactory ends.

The book starts with only the assumption that readers are somewhat aware of the notion that understandings of nature are coloured by and have incredible effects on the ways in which the world gets to be made (the background to this argument is spelled out clearly in Castree, 2005). The argument then builds, progressively. Readers may want to start at the beginning, get a flavour of the argument and then move at a speed that suits them through the book. The book is divided into two main sections, moving from a description of geographies of nature (or an argument on what they are and what they are not), to a discussion of how and why they matter. The two sections overlap. There are discussions in Part I that make strong cases for why it is that geographies of nature matter, just as there are dicussions in Part II that elaborate upon the spatial possibilities introduced in Part I.

The chapters are of varying lengths, and are sometimes written with slightly different styles, all of which is partly dictated by the subject matter and by the degree of introductory material that I have included. There are boxes or intermissions. Sometimes these are meant to stand alone and offer easily locatable places to gain a quick idea of a key point or style of thinking. At other times they can be used as side shoots from the main argument of the chapter, marking a change of pace and key, an example or case study. Their aim is to assist in the understanding of imaginative resources that a reader can bring to the main argument. Finally, each chapter contains a short list for background and for further reading. If you find a chapter particularly perplexing it may be worth looking at some of the background readings before carrying on. If you find the chapter is insufficient, or you would like more depth or empirical detail, then the further readings would be the place to start.

Part 1

What are Geographies of Nature?

Figure I.1   A small woodland in Greater Manchester, England

Are spaces for nature self-contained, sealed areas from which all trace of people has been banished? In Manchester, the city in England where I grew up, I remember that most woodlands were strongly fenced off, with warning signs nailed onto the trees saying, 'No trespassing', 'Keep out', and so on (Figure I.1). I ignored the signs, as did most children. The site in Figure I.1 is fenced off from the public, it's a local state-owned 'private' woodland. The positive side is that it has remained a woodland for as long as I can remember, the negative side is that most people cannot access it. Manchester City Council still has a policy of keeping nature reserves and people apart, fearing that people will interfere with wildlife. Its spaces for nature are, in theory and to some extent in practice, people-less. When applied to bigger areas, like, for example, wildlife reserves in Kenya, the term that is sometimes used for this kind of spatial practice is fortress conservation(Adams and Mulligan, 2003). The fortifications, which include guns, police and permits as well as fences, keep the world of people and the world of nature apart. And yet, when you start to observe these places you quickly note that not only are there numerous surreptitious border crossings (ranging from the rather innocuous fence climbing of my youth to the poaching parties that threaten tiger reserves in India), there are also lots of other crossings. Wildlife officers and volunteers enter the woodlands to clear sycamore saplings, brambles and holly under bush – all in order to maintain the habitat. In the larger projects and parks in India and Keyna, wildlife police, tourists, farmers, children, conservationists, scientists, animals, plants, remote sensing devices and animal medicines all pass through the parks. Meanwhile, fortress nature has long since been a contested practice. There are ongoing arguments over the best way to conserve nature – should people and wildlife be kept apart, or is it better (more realistic, more democratic?) to work towards the *in situ* co-presence of people and nature?

Both discourses of community participation and sustainable development have been mobilized to undermine what is sometimes regarded as the imperial practice of fortress nature. Whatever the answer, the point is that spatially things are not quite so pure and not so singular. Rather than watertight containers, spaces for nature are more permeable and multiple matters. So how do we think such spaces? This part of the book discusses some possibilities. In Chapter 1, I expand on the (im)possibilities for pure, sealed spaces. The focus is on spatial practices of conservation. The question is raised that, perhaps, given this porosity, is it that there are no spaces for nature, other than in our imperial imaginations? In Chapter 2, I discuss the possibility that nature exists more in human imaginations than on the ground. I look at some of the history of nature, focusing on changing understandings of evolution. In Chapter 3, I introduce a third type of spatial practice, that of enactment. The aim here is to use a number of examples, but mainly ones drawn from understandings of disease transmission, and specifically Bovine Spongiform Encephalopathy (BSE or mad cow disease), to explore the spatial multiplicity of nature. In Chapters 4 and 5, I build some more specificity into this discussion of enactment. In Chapter 4, I investigate common metaphors used to describe naturecultures, including interaction and hybrids. In Chapter 5, I take this forward to a discussion of nature and difference. By the end of this part of the book the aim will have been to suggest that nature is practised in ways that are spatially multiple. In Part II, the empirical practicalities of geographies of nature, how and why they matter, become the focus.

Nature's reality

Introduction

How do we think about and 'do' nature? And what does this mean for the ways in which we spatialize nature? In this chapter I want to explore and make some preliminary judgements upon three possibilities. We can sketch them quickly:

1   Nature as an independent state (but threatened by invasion)
    The first possibility is that we understand nature as something that is distinct from, absolutely separate to, the social world (Figure 1.1). Nature is another country, or is a part of ancient history, or buried deep in our make-up. It follows that Nature is real, 'out there'. 'Out there' meaning beyond us, or perhaps outside the 'in here' of our minds (so out there can include parts of our human bodies, those parts that are subject to natural urges, rhythms and involuntary movements).

    It may also follow that there is little of this nature left – for the social world is spreading, present as much in Antarctica as it is in our hormones. For most of the planet's inhabitants and history, 'in here' has had little or

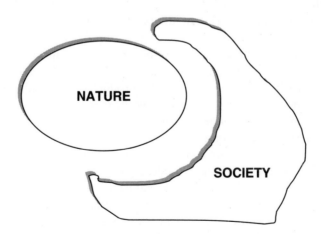

Figure 1.1   Nature as independent state (threatened by invasion). Nature and Society are separate spaces, but Nature is about to be or has been engulfed by Society.

no bearing on the workings of out there. Nature has gone on regardless of human imagination, dreams and schemes. Up until the agricultural, scientific and industrial revolutions of the last millennium, Nature out there was still much the same as it was when humans had barely started to scratch the surface of the planet. More recently this pure unadulterated Nature has become increasingly polluted in some form or other by human processes. The pollution takes at least two forms. First, there is the mixing of forms. Artificial molecules turn up in Antarctica. Second, there is the march of a form of rationality that sees the world as standing reserve, as of value only for human ends. Both mark the death of nature as Carolyn Merchant called it (1990), or the *end of nature* as McKibben (2003) termed this state of affairs. Not only is nature denuded, humans also suffer through an invasion of their own tissues but also through the repercussions of treating nature as an object to be governed.

In some form or another, this is probably the most common version of nature in Western societies. It informs many types of environmentalism, from the triumphalism of human mastery over nature to Western versions of stewardship and even some deeper green philosophies where nature needs saving from humankind, and humankind from itself. (The literature is vast but two of the best books remain Glacken, 1967 and O'Riordan, 1976).

2　Nature as a dependent colony, a holiday home
The second possibility regards nature as mainly, if not wholly, the product of human imagination. It is an idea. What is understood as natural is nothing but a product of the ways in which people order the world. Nature is ideological. It is socially contrived, produced by people and their value systems, political systems, cultural sensibilities. If there is reality, then that reality is social (Figure 1.2). Out there can be explained by in here. Nature, in this version of affairs, is a comforting illusion, or even a trick that people use to convince others of the faultlessness of their arguments ('it's natural that we do this, there's no point trying to change what's natural'). When we are told that the English Lake District or Niagara Falls are largely artifice, the product of hundreds of years of farming, design, literary and visual work, that they are ways of seeing rather than natural wonders, then we are starting to argue that what is taken to be natural in some quarters is, on the contrary, social all the way down. Likewise, when we contest fixed sexual identities, we're unsettling the fixity and conservatism of an ideological and already always political nature. Contesting the ideology of nature is often attractive politically, especially for a political project that is interested in gaining freedoms, or opposing those who would constrain liberties.

3　Nature is enacted (a co-production)
The third possibility I want to consider is perhaps the hardest and we will have to work at the spatial imagery. It suggests that nature and society

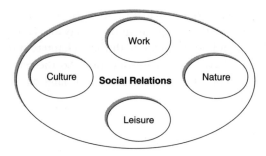

Figure 1.2 Nature as dependent state. Nature is but one of many categories that emerge from and exist within the realm of human actions and orderings. It is therefore dependent on and not prior to social relations.

make one another (so thus are not independent), but aren't necessarily reducible to one another (so thus are not strictly dependent). This is more difficult, but the basic argument will be that society and nature need not be considered as a zero-sum game. In other words, we do not need to think of a set amount of nature which is progressively eroded as society expands. Rather, the more activity there is in one, the more we might expect from the other.

This might be a more radical and interesting way of understanding nature, and one that this book is in part an attempt to elaborate upon. It's radical because it might well change the ways in which we attempt to practise nature. So, for example, nature conservation might well be a different practice once we view nature as neither totally independent of, nor totally dependent on, social worlds. It's also the least intuitive version of nature and requires us to do the most work.

Three possibilities, each of which has numerous variations and possible trajectories, and which will need a certain amount of teasing apart. You might, as we expand on each of these possibilities, become adept at spotting them in action (indeed, they are at work, practised in all manner of situations). But before we go on, I want to add that they are often mixed together in the same setting, making it more difficult to attach them as labels to organizations, people, or modes of thinking than might be supposed. My hope in raising the third possibility is not necessarily to call for some kind of absolute clarity. Rather, it is to suggest that where nature is concerned, things are often unclear, or not as clear as they seem. Our question then becomes how do we proceed, and proceed well, when clarity is always accompanied by murkiness. But before we get into these debates, it will be useful to use this three-part taxonomy, our three possibilities, to discuss how nature is mobilized in various settings. We will look in more detail at each in turn. In this chapter we will look at the possibility of nature as an independent entity (or, more accurately, its impossibility as I provide a critical review). In Chapter 2, we will explore nature as something dependent on society and culture. Again, the tone is largely critical, and, in being so, both these chapters start to trace the other possibility that I have called co-production. So the remainder of this chapter and the next involve laying groundwork for later chapters.

Nature out there

In our everyday language, we tend to treat nature and society as separate entities. If something is social, then almost by definition it can't be natural. And if something is described as natural, then it is unlikely to have much to do with society. So, for example, when we describe a landscape as 'natural' we often mean to suggest that it is undeveloped, untouched and that the social or human-made world is largely absent. But such a view, attractive and seductive though it can be for some, is often difficult to sustain. William Cronon, in a landmark essay entitled 'The trouble with wilderness' (1996a), launches a critique of this independent state version of nature, one that he argues has recently re-emerged in relation to the ways in which biodiversity and its conservation are imagined:

> The convergence of wilderness values with concerns about biological diversity and endangered species has helped produce a deep fascination for remote ecosystems, where it is easier to imagine that nature might somehow be 'left alone' to flourish by its own pristine devices. The classic example is the tropical rainforest, which since the 1970s has become the most powerful modern icon of unfallen, sacred land – a veritable garden of Eden [Figure 1.3] – for many Americans and Europeans. And yet protecting the rainforest in the eyes of First World environmentalists all too often means protecting it from the people who live there. Those who seek to preserve such 'wilderness' from the activities of native people run the risk of reproducing the same tragedy – being forceably removed from an ancient home – that befell American Indians. Third World countries face massive environmental problems and deep social conflicts, but these are not likely to be solved by a cultural myth that encourages us to 'preserve' peopleless landscapes that have not existed in such places for millennia. At its worst, as environmentalists are beginning to realize, exporting American notions of wilderness in this way can become an unthinking and self-defeating form of cultural imperialism. (Cronon, 1996a: 81–2)

One way in which nature independent gets done is, then, to expel all 'invaders' no matter how long they have been there, and no matter that they had a role in creating this landscape in the first place. It is worth reflecting too that these people were once simply labelled as part of nature, at a time when the separate continent of nature was not thought worthy of saving. As a part

Figure 1.3   Rainforest as Eden, 'La Forêt du Bresil', Johan Moritz
Rugendas

of nature, the people living there were often treated as unworthy of respect,
rights or political representation – a racism buttressed by naturalism. So
whether part of nature or not, people living in the continent called nature
have been anything but respected for their roles in ecological productions.

The message from Cronon and other environmental historians is clear. So-
called wilderness areas are peopled, have histories and geographies, and in
being so are in some way or another social as well as natural productions. In
a similar vein, forested and non-forested lands on the African continent are
as Fairhead and Leach (1998) have demonstrated, similarly peopled, and are
in fact co-produced landscapes, landscapes where people have had a hand in

developing the characteristic flora and fauna. Likewise, wild animals living in Kenya, so often visited by western tourists in search of the wonder and spectacle of nature-independent, are in some sense there on account of co-habitation with people (Thompson, 2002; Western et al., 1994). The list could be extended, but the point is made that what might look natural or wild to a western metropolitan eye is already mixed up with human worlds. To think otherwise and thereby to act otherwise (see Box 1.1) is to potentially do great damage to those people and to the landscapes, plants and animals that they have helped to make (and that have helped to make them).

Box 1.1   Thinking and acting

There's an unfortunate tendency to imagine that thinking and acting are either *unrelated* or *only related in certain ways*. In the first case, it is common to say that actions speak louder than words. We also often say that people think one thing but often do another. And the power of thought is weak compared to the power of bulldozers. Thinking seems harmless enough, compared, for example, to the violence that can be done with other tools. The phrase 'sticks and stones can hurt me but names never will' is something that many learn as a means to cope with the evil thoughts of others that in the end, we are taught, matter little. But as any child knows, names and thoughts are incredibly powerful and hurtful (a matter that feminist literary theorists like Judith Butler (1997) have usefully demonstrated). Thought matters, and can have effects. So thinking and acting are related. The way we think has repercussions. It follows that the way we think about something or represent a thing or an issue often shapes the way we enact it. If I think that wilderness is a people-less space, then I might feel the need to keep it that way. On the other hand, if historians convince me that this has not been the case, then I might think of ways to enact different kinds of wilderness, ones where people cohabit with wildlife. So thinking and acting are related to one another and it is not useful to make a hard and fast distinction between thought and action. An important adjunct to this argument is that even while this is often accepted, we still tend to assume that thinking and action are related in particular ways. So we might also note that it isn't simply that thoughts have effects, and are therefore important and powerful (as important and powerful as hammers and chainsaws). This would assume a cognitivist account, a linear narrative that first we think and then we act. This clearly is not the case. Actions and thoughts are not easily placed in such a sequence. Being frightened, for example, and running away or tensing muscles, may come before any sensation or experience of fright. Indeed, the release of adrenaline and

(Continued)

the reflexions of muscles are prior to thought. So actions shape thinking as much as thinking shapes action. We act to think. In doing so the world enters our thoughts (just as our bodies enter our brains through signals from nerves and muscles). We will have more to say on this later in the book, but it will suffice to say for now that thoughts are not made prior to action, and it is not a matter of some being enacted while others remain just thoughts. Rather, the world is the homeland of our thoughts (Ingold, 2000). Some thought-action assemblages will perform themselves more effectively than others, but their efficacy is not a matter of their purity or their neat sequencing – on the contrary, it is the more entangled, mixed-up thought-action assemblages that affect change in the world. So even though we can say thoughts are powerful, this does not suggest some form of crude idealism (whereby a state of affairs can be wished into existence). Thoughts are already of the world. Their agency, their ability to enact something, will already be interconnected or entangled with all manner of materials, tools, and others of all shapes and sizes.

This is an area of serious debate for it impinges on the ways in which natures are enacted, or practised. The important point to note is that Cronon and others are arguing that it's the idea of wilderness, as a people-less place, that threatens livelihoods and landscapes. So conservation for these authors is not necessarily about reducing the impact of people, it is about conserving some kinds of impacts, or disturbances, and viewing the space of wilderness not simply as a bounded territory but as a *collection* of effects, many of which connect to other places and times. The shape of the collective becomes a matter for political work (rather than a pre-ordained end, in the name of which all manner of atrocities of purification can be committed).

In addition to the observation that wilderness is not a people-less place, a territory independent from human societies, there is the point that by labelling and looking at wilderness, it becomes a social matter. The fact that wild-scapes are valued, pictured, imagined, visited, monitored and measured also starts to unsettle the sense that they exist purely and simply elsewhere, divorced from and entirely separate to human and social worlds. All of these practices in some way or another touch the worlds that they seek to value, measure, picture, and so on. So even if wildernesses are 'successfully' depopulated, their 'enframing' as objects to value, to view and maintain is itself already a form of inhabitation. It too has effects. I return to this point in more detail in the next section.

Hopefully, you are starting to be convinced that 'independence' is a rather dangerous metaphor (even more so when we link it to ideas of purity). But even so, even if landscapes, animals and perhaps species of plant are tangled up with humans, so that their histories and futures are intertwined and depend upon one another, might it be the case that Nature still exists, separate to people, at a more fundamental level? Cronon, in places, and other environmental historians argue as much (see, for example, Worster, 1988). So we need to go a bit further in order to explore the independence of nature, its apparent out there-ness. One place to start is an ancient distinction between natural objects and natural forces.

Two species of Nature

The difference between talking about nature as an object (a scene, an animal, etc.) and nature as a force (or process) has been around since at least the Middle Ages in Europe. The two ways of approaching a discussion of nature are sometimes expressed in the terms *natura naturata* and *natura naturans*. The first of these, *natura naturata*, is used to describe the products of nature that we can observe with our senses (trees, mountains, animals, microorganisms, wind, and so on). The second, *natura naturans,* is the so-called invisible or less tangible force of nature (see Adam, 1997: 30).

Following Cronon and others, we have already suggested that what might appear to western eyes as natural objects, as *natura naturata,* turn out to be tangled up with humans, their material effects, their ways of seeing, their ways of ordering the world. But what about natural processes, *natura naturans*, surely, these are what they seem? Even if people have inhabited wilderness areas for millennia, even if elephants and people walk together in Amboseli, even if forests exist partially as a result of human actions, even if organically produced tomatoes are still co-productions, surely this is a surface level phenomenon, and underneath it all are the biochemical processes, and geological upheavals that range from sub-atomic to interplanetary spatial scales, and from nano-seconds to billions of years. Surely these are elsewhere to the social, out there or deep in there, outer and inner spaces, unaffected by and indifferent to people with their axes, arguments and aesthetics?

Here are two arguments that suggest otherwise. The first argument is the suggestion that natural processes are now so polluted and mixed in with contemporary society that they have ceased being very natural at all. The sociologist Barbara Adam makes this kind of argument:

> Animals grazing peacefully on a hillside, waves lapping gently up the pebble beach, a pine forest whistling in a storm, a river bursting its banks, a hurricane tossing houses and cars in the air like play-things, a bush fire raging out of control – all are images of nature, some idyllic, others threatening. Can we be sure, however, that this is nature in the conventional meaning of the word, that is the result of forces uncontaminated by human activity and production? What becomes of this understanding of nature when those grazing animals are

contaminated with radiation or suffering from BSE, when the pine forest
(a monoculture, likely to have been planted during the last century) is suffering
from the effects of acid rain, when the flooding is due to agricultural practices
that have led to oversilting, when the extreme weather conditions are linked to
global warming, and when the bush fire and the scale of its damage have been
facilitated and exacerbated by human actions? During this century it has
become increasingly difficult to sustain the division between nature and culture.
When even the stratosphere is affected by the industrial way of life, when the
sun is turned from source of health and well-being to health-hazard and danger, when the air we breathe causes respiratory diseases and allergies, when the
traditionally conceived untamed, raw power of nature is so extensively influenced by human action then the traditional separation between nature sand
human culture collapses. (Adam, 1997: 26–7)

For Adam, this merging of nature and society is linked to a specific period of
human history. As more and more of nature became affected by the 'industrial way of life', so nature became less recognizable as a pure category in
the nineteenth and twentieth centuries. In making this argument, Adam was
drawing on similar resources to those of Bill McKibben, who famously
declared the end of nature in 1989 (McKibben, 2003). McKibben drew attention to what he saw as two reversals in the relative status of social and
natural worlds. The first of these referred to nature's time, the second to its
space.

McKibben suggested that contrary to the notion that nature moved in
another time to people, and that there was an infinite slowness to natural
processes that made human time scales (their life times and their parliamen-
tary times) look either irrelevant or at least puny, the world was now so
infected with the human organism that it was speeding up. It was changing in
rapid and dangerous ways. Climate change was the paradigmatic but not the
only example cited. Related to this was another reversal. People had become
used to viewing themselves as small players in a large world, but, as has now
become a common feature of globalizing stories, people, or at least some
people, are now large (in number, in effect, in reach) with the result that the
world had become a good deal smaller. In sum, then, nature, whether characterized as objects or as forces and processes, was infected with humans, so
much so that its great times and spaces had been overridden. Nature was now
small and fast running out. The independent state of nature (which *had*
existed, at least for McKibben) was at an end.

The second way of making the argument that natural processes are not
purely natural draws upon an important form of analysis in social science
thinking for which I will use the term ontological politics (see Mol, 1999).
The body of work is complex, varied and easily misunderstood (for an excellent account, see Law, 2004a). It informs a good deal of the arguments in this
book but let me make a start by suggesting two ways in which it can be
thought. First, the way that nature is viewed, understood, made sense of,

written about, pictured and used is in part a result of the position or place within which viewers find themselves. As the geographer David Livingstone (2003) has put it, science and knowledge have geographies (as well as histories). What is made present in a field, a laboratory, a research article, a poem is in part a product of the sedimented practices that inhere in the ways in which laboratories and languages, to name but two, work. Place and space matter. The all-seeing, god-like view, divorced from all the messiness of worldly matters, is a trick, a god trick (this is the famous view from nowhere, a device rendered in landscape painting in the seventeenth century, see Alpers, 1989). It is a view that was taken up in the development of scientific practice and in particular at a time when objective or viewer-independent accounts of the world were deemed to be important (for a classic account of this moment in Western science and politics, see Shapin and Schaffer, 1985). The obliteration of the practice of observation and the resulting focus only on the object being observed produced a strange account of the world where such objects became independent of their human relations, and could therefore be treated as lonely matters of fact.

Against this purification of objects and facts, it may well be more useful to think about views from somewhere, situated knowledges, or partial perspectives (Haraway, 1991b). Viewers are embodied, passionate, political, social, temporal and spatial, which is not the same as saying they are local. They are spatialized, connected and disconnected in varying ways to others and lots of elsewheres (Massey, 1999). So this is not an opposition between local knowledges and global science. Or between particular views and universal laws. Rather, all knowledges are situated and more or less connected or connectable in order to make more or less consistent spaces for that knowledge. What follows from this is that any understanding of nature as thing or force (*natura naturata, natura naturans*) is infected by all those things, allies, journeys, languages, loves, funds, and so on that go to make that understanding possible. That, in any case, is one argument, but we need a second one. Otherwise, it might sound as though 'location', 'perspective' or 'social context' is the only thing that matters.

Second, it is important to add here that this is not simply a matter or an issue of epistemology (or simply put, the way in which the world comes into view). It is not simply that there are multiple ways of viewing the world. To say so would not do much work if we are intent on unpacking the purity of *natura naturans*. Even if it is acknowledged that there are many viewpoints on an object, some more or less polluted by politics, emotions and other kinds of bias, this might only delay the moment when nature proper is declared. Once we have cleared away all the bad views, we can get to the truth. Once we have progressed from all the quaint, old views, we can truly consider ourselves enlightened by the one true version of affairs. So if we stopped at epistemology we could get into the long and painful history of epistemological politics, charting some of the battles that have been waged over different views of what are ostensibly the same thing (and then arguing over the grounds on which we can demonstrate or agree that one of these views is more accurate than the others). The argument in this book is more demanding than this notion that there a number of possible perspectives on the same thing, and that either we can decide which is the best (often that which approximates closest to the god-trick or disembodied view from nowhere – traditionally a view from an objective, white-coated, male, emotion-free science – or sometimes reversing the polarity and valorizing those who are deemed to be closer to nature, historically often essentialized subjects called woman, indigeneous, and so on), or agree to differ and say that all kinds of beliefs are possible (a form of relativism, or, in political terms, liberal pluralism, accepting the views of a plurality of positions as of equal validity).

Box 1.2 From the view from nowhere to the view from whereabouts

The development of a view from nowhere in western epistemology foregrounded the viewed object and made viewing practices invisible. It is a view that clearly parallels and supports an independent nature – for the object can stand by itself and is independent to the processes of bringing it into a frame of reference or view.

Citing the impossibility of viewing without having some kind of interference between object and subject, social scientists tend to talk of a *view from somewhere*. Instead of the disembodied and invisible god's eye view, we have an embodied practitioner with all their equipment, stories, funding agencies, language, and so on.

The tendency, however, in stating that there is always a view from somewhere can be (and this was certainly not Haraway's intention) to place the viewer in a fairly static field (defined by their gender, their language, their time or their economics). So there may be a better way of getting at the complex practices of viewing.

One place to start is Merleau Ponty's very different sense of a view from nowhere (Merleau-Ponty, 1962). In his case, rather than suggesting the bird's eye view was one that effaced practice, he focused on the complex practices involved in generating more than one view of something, and the process then trying to piece those different practices and views together. Merleau-Ponty used the example of a house which can be viewed from a variety of somewheres – inside, from the landing, from the road, above if you were able to fly over it or could climb to some other vantage point. But

(Continued)

(Continued)

what Merleau Ponty is interested in is not the proliferation of possible views, but the way in which these views are combined together to form a view from nowhere. Tim Ingold summarizes this nicely.

> The house is progressively disclosed to me as I move around and about, and in and out, not as the sum of a very large number of images, arrayed in memory like frames of a reel of film, but as the envelope of a continually changing perspectival structure. Observation, Merleau-Ponty claims, consists not in having a fixed point of view on the object, but in 'varying the point of view while keeping the object fixed' (1962: 91). Thus the house is not seen from somewhere but from nowhere – or rather from everywhere. (Ingold, 2000: 226)

This seems useful, but there are two problems. First, the *view from everywhere* sounds too totalizing. Even though views can combine, the result is surely more partial than that. Second, the focus seems to remain thoroughly human, and the 'fixed' objects seem to be waiting there passively to be sensed in this albeit more active way (see also Chapters 4 and 5 for a discussion of this aspect of phenomenology's anthropocentrism). So we need a view that doesn't hold the object fixed, but allows some movement of subjects and objects. Later chapters will expand on this multiplicity, but for the moment I want to suggest that we use the term '*the view from whereabouts*' to figure two things. First, there is likely to be more than one practice involved in making a view. Second, the thing being viewed will not be fixed but can also move and alter, so that its location, like that of the viewer, may be approximate, or whereabouts.

The argument that I want to pursue here draws on the work of Haraway (1991b), Latour (1999; 2004b), Mol (2002), Law (2004a) and others, all of whom work in a loosely defined or gathered field called science, technology and society (STS). Their take is roughly as follows; it is not simply that there are many views on the same thing, it is rather that views and things depend on one another (see Box 1.2). Views enact things differently (and actions can alter views). This may sound counter-intuitive. But let's go back to the case of tropical and subtropical forests. Cronon (1996a) and Fairhead and Leach (1998) all make the case that first world conservationists have viewed forests as peopleless places, the inhabitation of which causes a threat to the vegetation and wildlife that currently live there:

Not only did the development of scientific ideas about West African forests have its own complex intellectual history and sociology, in which certain theories or debates were able to rise to the exclusion of others. But also, and crucially, these views dovetailed with the administrative and political concerns of the institutions with which they co-evolved in a process of mutual shaping. Ideas about forest-climate equilibria, or the functioning of relatively stable forest ecosystems, for instance, fed directly into a conceptual framework and set of scientific practices for conservation, which was about external control. (Fairhead and Leach, 1998: 189)

The point here is that views can have effects. To be sure it takes the right circumstances and some neat joining together of knowledges and ways of thinking (in this case, a dovetailing of colonial rule and imperial knowledge). There's a performance of a god trick too, as scientific expertise is presented as placeless, or better, applicable everywhere, in order that it can take over the running of forestry practices (and thereby displace other practices, regarded in this case as non-expert, local and unscientific).

Anthropologists and indigenous forest dwellers, meanwhile, view the forest differently. We don't have to romanticize these views or even suggest that they are somehow more natural, to nevertheless suggest that they enact the forest differently to the scientific ecological view. Indeed, rather than seeing the forest as a delicately balanced ecosystem, Fairhead and Leach suggest that forests and forest margins are lived as dynamic, changeable places, where adaptability is key to survival and where boundaries between forest and savanna are in flux.

On the one hand you could characterize this situation as two views on what is essentially the same thing. But another argument would be to say that the forest is different depending on which one we listen to. There may be some similarities between the two views and the two objects that they help to shape, but there are also some pretty big differences (one is peopled, the other is or 'should' be people-less; one depends on people, the other depends on their being made to leave; one is accorded a natural balance, the other is part of a dynamic of continual disturbance). In epistemological terms, we want to be able to decide who is right and who is wrong. We could subject them to the same trials of knowledge, the winner being whoever has the dominant vocabulary of causation. But the sociology of science has taught us something else – it isn't the power of argument alone that wins. Those who can make the world in the image of their arguments, who can, as Bruno Latour famously put it, 'make of the outside a world inside which facts and machines can survive' (Latour, 1987: 251) are the ones who carry the day. So in disputes it is not simply epistemology that matters, it is also necessarily a question of which side is building the more robust networks, who is turning arguments into actions (see Box 1.3).

Box 1.3   From single reality, to multiple realities, via discourses and associations

How do things get done, how do they get made? After divine ordination came rationality and nature. Things were done this way because that is the way things are, in nature. Again, nature independent looms large in this kind of story. But what if rationality, nature, and so on are not so fixed and are also in the making? For many in the social sciences, the answer lies not in fixed logic or the timeless order of things but in the power of discourses, or linguistic and material arrangements which convince others of the importance of their arguments. This is, in any case, a common reading of early Foucault and is somewhat present in the notion that ideas develop in historical contexts and then shape the way things happen. (The quotation on page 19 from Fairhead and Leach provides an exemplary case in point.) A good deal of actor network theory takes a different approach, arguing that it is more than ideas that make things happen. For authors like Latour, in his famous case study of the Pasteurization of France (Latour, 1988), it was not ideas or logic that produced change in the French countryside but the hard practical work of demonstrating the advantages of the method on farms, of enrolling farmers onto the programme, of solving problems in the field or making the world outside one where the world inside the laboratory could work. Making things work was therefore a practical and material matter of association, not one of convincing others through logic and ideas (see also Mol, 2002: 61–71). A subsequent step in this shift from human ideas to the practicalities of things would be to ask, what if more than one thing was being enacted simultaneously (in our case, more than one forest)? Is it simply a case of one forest becoming associated and the others dying out, or can they coexist, inhabit more than one network and even work in other kinds of space? Is there more than one forest, and more generally, more than one space for nature? The inspiration here is in the work of Mol, Law and others (see Law, 2002; Mol, 2002). We will come back to this issue of the multiple in subsequent chapters.

So both versions of the sustainable forest are more than ways of seeing, they are ways of intervening and engaging, and they perform their objects differently. Another way of saying this is that they are interventions in the making of forest. We will look at many more examples of this enaction of knowledge within the book, but the main issue to note for the moment is that we are starting to mix questions of epistemology (ways of looking, what is known about something) with questions of ontology (ways of being, or enacting what is). And because we are suggesting that things are not settled, timeless or given, then these realities are in the process of being made. We're now starting to open a politics not simply of who has the best view, but which is the more effective and active form of world-making. This is what Annemarie Mol (1999) has referred to as ontological politics. There is, in this case, not one forest which must be secured to the exclusion of all other versions, but possible forests that can be enacted differently, depending on, in this case, both the knowledge *and* the politics of forest inhabitation. Meanwhile, just as there is more than one forest, we could also add that these versions overlap as well as pull in different directions. Another way of saying this is that the number of forests is not unlimited or infinite. They are multiple but also connected (see Box 1.3). The possibilities are not endless, but neither is unity or absolute agreement between all the people and all the things necessarily possible or desirable. A mantra of ontological politics is that there is always *more than one but less than many* forest/s, disease/s, city/cities, aeroplane/s, water vole/s. That's the exciting if challenging aspect of the politics that inhabits the pages of this book. It is neither a politics that is necessarily subservient to Science which is asked to adjudicate on all matters of substance, in order to find the one true version of affairs (back then to epistemology wars). Nor is it a politics that is happy to let anything go, to accept as many truths as there are parties, to think that all these versions of forest can coexist happily if only we could agree to differ. So neither uni-verse (one world) nor pluriverse (many worlds) will do. In philosophical terms, neither monism, dualism or pluralism will do. The numbers are going to be more difficult to imagine. Fractions rather than fragments are needed. A term that is commonly used is multiple, which in this case is not equivalent to plural. So when Annemarie Mol talks about the body multiple (Mol, 2002), for example, it is not to suggest that there are endless ways in which a body can be viewed. It is simply to underline that there will never be a single body which dictates what happens next. There are multiple versions, realities, being performed, which are not mutually exclusive, as they affect one another. The ways in which this multiplicity and connectivity are dealt with become the subject of various forms of dealing with difference, including negotiation, indifference, struggle, and so on.

The point has been to suggest that it has become difficult to sustain a view of nature as an independent state, not simply because of human expansion into all corners, into the tiniest and the largest of earthly matters. It is also that, first, all these matters are viewed and made sense of in ways that cannot be totally divorced from their times and spaces, and, second, in *making* a view (and I should emphasize making), the viewed thing is also being made, it can be affected by the very process of being attended to. To be sure, the degree of effect may be variable, and it is not something that humans do on their own – ontological politics involves trees, elephants, soils, ants, mountains, water, ocean currents as much as and often more than human beings. But that's part of the task, to work out a politics that is more than human (Whatmore, 2004), that is attendant to the mixtures and separations that make things and make nature.

Conclusion

Maybe you are convinced that some forms of nature, like landscapes, can be co-productions, but highly doubtful as to whether the smaller stuff and the really big stuff, the things that look indifferent to humans, can be anything but independent. You may now be partly convinced that independence is not always as clear as can sometimes be suggested. The quote from Barbara Adam is a useful reminder of the depth to which human and nonhuman worlds have become enmeshed. But you probably still have a lingering doubt over the argument that independence has had its day, even when the intricate arguments of ontological politics are introduced. If you do have doubts, then these are well placed. There is much more work to be done before we can dispense with independence and find other ways of understanding nature. Nature's reality and indifference to humans *are* an issue and one that I start to pick up again in later chapters. But for now we need to look at the second possibility, that of dependence. This is the subject of Chapter 2.

Here are some preliminary conclusions to the argument so far:

Firstly it's difficult to find pure nature.

Secondly even distinguishing form and process doesn't necessarily help as humans have managed to infiltrate most aspects of the world, and even where this can be doubted, their engagement with an object is also part of an intervention (no matter how insignificant this may seem) in that object's world.

Thirdly we need to attend to the ways in which nature is addressed as both real and made. To do this we need a more subtle spatial imagination than 'independence threatened with invasion'.

Background reading

Noel Castree's (2005) book *Nature* is a clear review of geographers' engagement with nature, and includes a useful review of the power of ideas in shaping how worlds are made. Bill Cronon's (1996a) wonderful essay remains the best introduction to thinking through the idea of wilderness.

Further reading

James Fairhead and Melissa Leach's (1998) work on African forests and forestry is a detailed work on the role of people in making natural landscapes. Barabara Adams' (1997) work extends the current argument to consider time in more detail. John Law (2004a) provides an extremely clear introduction to the sociology of science and to ontological politics.

2   The thought of Nature

In Chapter 1, I suggested that nature's independence is, at the very least, a fragile state of affairs. But what if we go further and make the argument that, rather than being an independent state, nature is totally dependent on humans – it is constructed all the way down? Its out there-ness is a trick of the mind. In the last chapter we called this second possibility nature dependent. While we will want to unsettle such a spatiality of total dependence (rather as we have started to do for total independence), it is nevertheless useful to explore just why such a view crops up in the first place. The case study of evolutionary thought helps us to explore the usefulness of being able to criticize independent views of nature and claim instead that they are constructions that owe a good deal to the social worlds in which they are made.

The idea that the evolution of living things occurs as a result of competition for scarce resources (a process referred to as natural selection and derived in part from Darwin's famous (1998) work, first published in 1859) is widely reproduced. To be sure, the idea of evolution is contested, with the most prominent challenges coming from Christian creationist groups. But on the whole evolution tends to get rolled out as a robust set of statements that refer to the workings of nature out there. It is an idea that seems in this sense to be timeless and universal. And yet, the idea of evolution, which in its simplest sense means change over time, has itself changed over time. In this chapter I want to dwell on a small part of evolution's history. The aim is to illustrate how 'nature-dependent' arguments work. They work, I will argue, by demonstrating how what seem to be pure descriptions of nature are in fact socially constructed natures, born not of unmediated contact with the outside world but of thoroughly social mediations of nature. To be clear, this is not the argument of this book, but it is useful to see why and how such arguments are made in order at least to flag up some of their problems. I will start by looking very briefly at how evolution tends to be presented in contemporary culture, and follow this with a slightly more in-depth look at the process by which Darwin's account was produced and reproduced in the nineteenth and twentieth centuries.

Figure 2.1   Detail from an advertisement for the car that has evolved

Evolution as competitive individualism

A series of adverts for cars that appeared in the British media in the 1990s depicted a new breed of car alongside an animal, normally a predator (an eagle and a wolf were prominent in the series). The by-line of the adverts suggested that, like the animal, the car in question had evolved. It was leaner, stronger, more powerful and more suited to today's driving environment. I want to use those adverts, loosely, to think about the kinds of ways in which evolution is imagined (Figure 2.1).

The car and the animal are both products (*natura naturata*) that have emerged from similar processes (*natura naturans*). Both have been formed by forces which are in some ways similar. The car is shaped by a competitive market – or economic forces. The animal's strength, brains, and agility have been formed through millennia of competitive selection – natural forces. In all of the advertisements a single animal was placed together with a single car. It is unsurprising to see a single car, perhaps, for the last thing that car makers want to show is a world crowded with cars. But perhaps we should be more surprised by an image of a single animal?

In the terms of the advert, it makes sense to view the individual animal as separate to its environs. In the process of natural selection it is the *individual* who is smart, strong, lean and swift who will be favoured. This is the 'survival of the fittest'. (One irony of this advertisement and the series of which it was part was that many of the animals featured (the golden eagle, the wolf) were listed as threatened species globally.) However, the abstraction of the individual seems strange when we consider that these animals are rarely 'alone'. Another way of representing wolves, for example, is to suggest that they are highly social animals, living in well-organized social groups. They maintain strict hierarchies based partly on strength but also on role. Importantly, they co-operate skilfully in hunting and share tasks of cub rearing. They also evolve with other species, their future entwined with those of their prey (herbivores of various kinds), their prey's prey (various plants) and others (notably carrion who benefit from wolves' tendency to leave a kill once they have eaten their fill).

Following this more collective account, evolution can be regarded as a more relational matter, where change is social and ecological and where it makes less sense to abstract individuals or individual species. So while natural selection can often be read as tending to imply that evolution occurs through competition for scarce resources between individuals of the same species (as well as between individuals of different species), there are plenty of examples of species co-operating. And these relations might not stop at the species boundaries. Assemblages of plants, animals, and non-living matter may co-evolve and produce opportunities and constraints for one another through all manner of relations including co-operation, symbiosis, parasitism, co-habitation, opportunism as well as competition (for an example, see Box 2.1).

Box 2.1   Interspecies relations

In Yellowstone National Park, USA, the re-introduction of wolves (*Canis Lupus*) in the mid-1990s resulted in the regeneration of riparian cottonwood habitats that have been in decline in the region for nearly 70 years (Figure 2.2). Wolves had been hunted to extinction in this part of North America by the 1930s. The connection between wolves and the habitat had not been made previously, but it turns out that wolves may be vital to the maintenance of cottonwoods through their predation on herbivores (whose own browsing on

(Continued)

(Continued)

young shoots reduces the capability of cottonwoods to reproduce) (Beschta, 2003). This is an example of what ecologists call a trophic cascade. 'When a top trophic level predator interacts with the next lower level herbivore and this interaction in turn alters or influences vegetation, a "trophic cascade" occurs' (Ripple and Beschta, 2003 : 300). This interaction can be both a function of reduced prey numbers but also shifts in the behaviour of prey as a result of the predation threat (indeed, the results of the ecologists' studies in Yellowstone suggest that it is a learned behaviour of elk (*Cervus elaphus*) who are avoiding river bank grazing owing to the reduced options for predator evasion, constrained as they are by water on one side, that is the main change). The authors of this work suggest that wolf extirpation produced relatively low risk winter grazing possibilities for elk which resulted in low stature trees with a reduced range. Extinction of the top level predator also coincided with 'a drastic decline in the northern range beaver (*Castor canadensis*) population' (ibid.: 301), possibly as a result of reduced foraging possibilities caused by habitat degradation in the early twentieth century.

Figure 2.2 Floodplain with sparse cottonwoods photographed in 1969. Regeneration of riparian cottonwood took place after the re-introduction of wolves.

The point in this chapter is not necessarily to suggest that evolution is one thing and not another (although clearly I am more sympathetic to the more ecological, complex interrelations that involve competition, co-operation, mutualism, and so on than the notion that it is competitive individuals that drive change). Rather, I am interested in suggesting that evolution is complex, and that there is more than one version of evolution to consider.

(a)

(b)

Figure 2.3   Natural images as indicative of social organization.
(a) Mosaic of a worker bee incorporated into Waterhouse's Manchester
Town Hall (1877) (b) Image of DNA used to convey the knowledge-base of
an organization, in this case, an open source software format.

One way to handle this multiplicity is to suggest that these views are all
generated in different social situations, and they are therefore coloured or
even conditioned by the social context within which they are made. They rely
on the language and ideas that we have available to us in our social worlds.
They use ideas about individuals, about competition, about information and
blueprints, about social behaviour and about co-operation. The language and
ideas that are used to identify, describe and explain the natural world are
influenced by the kinds of societies people live in, believe in and/or want to
secure. In the advert, the social world that is being sold to us at the same time
as the car is one of competition and individualism expressed through consumption. Elsewhere, other metaphors are used to link nature and society
together. In nineteenth-century industrial Manchester, for example, it was the
bee (Figure 2.3a) that was taken as the natural symbol of a city intent on foregrounding work, co-operation and municipal improvement. More recently it

is often DNA that is taken as the symbol of informed and knowledgeable organization (Figure 2.3b).

So nature is presented in different ways, but surely this is a secondary matter, applying to cultural products like adverts and buildings, and not to scientific texts? Culture is of course constructed, but what about nature itself as revealed through science?

Darwin, evolution and its impacts

The Origin of Species, the first of Darwin's more controversial works, was first published in 1859 (Darwin, 1998). It is a rich and complex text that can be read in various ways but the overwhelming message, and the one that was promoted by keen allies, like his 'bulldog' Thomas Henry Huxley (Desmond, 1998), and latched onto by mid to late nineteenth-century readers, was that nature was in the process of continuous, long-term change. And these changes, or transmutations, of animal and plant species were brought about through natural selection. The latter was largely the outcome of a brutal competition for scarce resources. Over many generations, ruthless competition resulted in the fittest individuals surviving to reproduce and pass on their traits. In brief, then, Darwin's work suggested that:

1. The natural world is not fixed, it is changing.
2. The natural world changes according to some specific rules and is ordered through competition.

Is this a view from nowhere, or did Darwin's situation, and that of his readers and those who would champion certain of his ideas, matter or have effects on the theory of natural selection?

The first thing to emphasize is that evolution was not a new idea. It was a concept that had been positively resisted by most established scientists in England in the first half of the nineteenth century. There were radical voices outside the establishment, notably in some of the Scottish universities and in post-revolutionary France, but on the whole these dissenters were ridiculed in England, a ridicule that was also an expression of fear. Evolutionists were feared because a challenge to the fixed order of nature was also a challenge to the fixed order of society. The Church, the Monarchy and the landed gentry gained some of their authority through the claim that their ascendancy in society was part of the natural order, ordained by God. Any attempt to disrupt that order would be heresy. And if anyone was in doubt that this social order was the way it was meant to be, then they only had to look at the way nature worked. The latter was generally thought of as a hierarchy of beings placed on earth by a divine will to live out a specific role. Suggesting otherwise, that change was the norm, might threaten the established social order as well as the natural order.

Second, Darwin's background made him sceptical of this divine order, but also fearful of its demise. He had had a privileged upbringing, and himself became a landlord and squire later in his life. But his family was in some ways non-conformist. They were part of an emerging professional class, and like the wealthy, industrialist Wedgwood family into which Charles had married, worshipped in a non-established church (on the Wedgewoods and non-conformist spaces, see Hetherington, 1997). Charles had also spent time at Edinburgh University where he was taught by radical evolutionists. But soon after dropping out of medical school, he went to Cambridge to study theology and developed a lasting affection for the establishment life of gentle science combined with Church of England doctrine. Indeed, his ambition had been to live a quiet life as a rural pastor, a life which would enable him to peacefully continue to pursue his passion – collecting beetles.

As Darwin began thinking seriously about the transformation of natural species, he was aware of the potential for evolutionary debates to destabilize not only ideas about the natural world, but also the social worlds within which he lived. This was not simply paranoia, Darwin had seen how arguments concerning the transmutation of species had been used by militant groups as a way of unsettling the 'natural' order of the established church and gentry. And he had witnessed how these militant thinkers had been treated. For example, in 1844, Emma Martin, a feminist campaigner, published her inflammatory pamphlet *Conversation on the Being of God*. In it she argued that evolution needed no creator, and through this argument she outlined a mandate for social change (Desmond and Moore, 1991: 316). She toured the country, enchanting audiences with her engaging speeches, but she was soon summoned to court for causing disturbances at churches and was hounded out of the towns and cities that she visited. Emma Martin was just one of an urban radical movement which was rehearsing revolutionary and reformist arguments in the politically unsettled Britain of the early to mid-nineteenth century. Darwin was more than concerned that his ideas would add fuel to their fire. But he was also dissatisfied at the conventions of scientific thought and its priggish strictures.

When Darwin did go public with the ideas and notes that he had been nurturing for over 20 years, his work was in part embraced by radical thinkers who were keen to find justification for the transformation of society. Indeed, Karl Marx and Frederick Engels applauded Darwin for managing to lift the veil off conservative science and reveal the true transformational character of the world. Yet the socialists, anarchists and revolutionaries found certain parts of Darwin's logic unacceptable. In particular, while they applauded the emphasis on transformation and change, they were appalled at the process through which this change was brought about.

If the concept of gradual change emerged in a society concerned with reforming the old structures but frightened of social revolution, then the mechanism for this change was even more remarkably tuned to the social world of which Darwin was part. Darwin was a wealthy man, with investments in land and now in railway companies. In many ways, he was concerned with maintaining a liberal economy in order to protect those investments. In this he was very different from the authors who were publishing rival theories on evolution at the time. These were often written by urban radicals concerned to better the lot of the poor and overturn the established order. They were writing from atheistic positions, concerned in part to destroy the privilege afforded to the Anglican Church. One article, on the 'Origin of man: Science versus theology' and published in the *London Investigator* in 1854, five years before Darwin published his theory, was anti-Creationist and suggested that the human species had evolved upwards. Meanwhile, like many other evolutionary theories of the time, this article emphasized co-operation as the means by which humans had ascended. These scientific stories were not about the fit and the rich annihilating the weak. They were critical of such elitist brutalism, and offered a more democratic, bottom-up theory of change (Desmond and Moore, 1991: 413).

Partly on account of his social position, Darwin was often scathing of this kind of unphilosophical and overtly political work on evolution. For similar reasons, he was also attracted to other ways of accounting for evolution that did not highlight co-operation. But it was his reading of Thomas Malthus's (1798) *Essay on Population* which helped him to construct another way of providing a mechanism for evolution (Malthus, 1992). The Reverend Malthus's main, and explicitly political, argument was that God had made the human species reproduce at a faster rate than was the case for its surroundings. The result was that there would be an endless cycle of population expansion, inevitably followed by scarcity, misery and retrenchment. The reason God had done this was to deny humans the ability for sloth. Humans could only achieve their full divine capacity through work and thus advance towards civilization.

Malthus's ideas were employed directly in Britain to argue for reform of the Poor Laws, and were used to argue against the extension of welfare to those who couldn't work. Welfare, Malthusian extremists would argue, only ensured the over-reproduction of a non-productive class of dependants. This would lead to a slothful population and would slow down the civilization process. This Malthusian understanding of population theory was and is highly contestable. Indeed, Marx and Engels along with many left-thinking writers of the past two hundred years spent a good deal of their time refuting the notion that there are fixed, physical limits to the size and affluence of a population. Even at the time that Darwin was formulating his theory of evolution, people attacked Malthus's work for its pessimism. People even doubted whether more food led to higher rates of reproduction. Arguments were made that suggested animal and plant species reproduced more in times of stress. People suggested that potting plants in rich soil could lower fertility. Extending this to human populations, they suggested that welfare and reduction of poverty were necessary to help to lower population growth – an argument that has parallels in contemporary debates over famine and economic development. Darwin was unconvinced by these anti-Malthusian arguments. He took from Malthus a doctrine of the struggle for existence. He experienced a London during the economic recession of 1840s, he followed the Crimean War in the newspapers, he drew upon knowledge of and investment in a growing industrial sector where factories were developing new forms of organization – the division of labour – which could improve economic performance (and seemed to do so during the 1850s). All of this conspired to convince Darwin of the struggle for existence, the need for specialization and the importance of competition.

When Darwin published his theory in 1859, it was received much more favourably than previous evolutionary theories. This was, in part, a result of Darwin's already well-established scientific reputation. He was a well-known author on such matters as geology and zoology. But it was also a result of the support that his theory seemed to give to the ideas concerning economic progress that were being touted by the wealthy classes. It was, after all, a theory which seemed to give legitimacy to the idea that competition, and free competition at that, was necessary in order for there to be evolution, or positive change. Now this was certainly bending Darwin's words – he was suspicious of suggesting that evolution was always synonymous with improvement or progress. Nevertheless, it was far from difficult to apply Darwin's ideas to British society and to the British Empire. The ideas had, as we have seen, partly been derived from those very spaces.

This is not to say that Darwin's work was accepted uncritically. Its Malthusian tones were heavily criticized by those commentators who were less convinced by the mechanisms suggested for change. Frederick Engels, for example, had this to say:

> The whole Darwinist theory of the struggle for existence simply transfers from society to living nature Hobbes' doctrine of *belum omnia contra omnes* [the war of all against all] and the bourgeois-economic doctrine of competition together with Malthus' theory of population. When this feat has been performed (and I question its absolute permissibility ... particularly as the Malthusian theory is concerned), the same theories are transferred back again from organic nature into history and it is now claimed that their validity as eternal laws of human society has been proved. (Quoted in Atkinson, 1991: 105)

Engels was saying at least two things. First, the knowledge that Darwin had constructed concerning the natural world had been influenced by a selection of social doctrines and beliefs of the time. The knowledge was, in Engels'

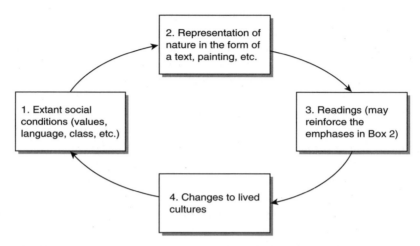

Figure 2.4   The construction of natural knowledge and its effects

opinion, political. Darwin had selected particular theories of divinely sanctioned social organization (including Malthus's doctrine of competition, but also Hobbes' liberal notion of politics) and used them to account for change in the natural world. Second, Engels went further to suggest that not only was Darwin's knowledge of the natural world influenced by the social milieu, it also had effects – it had an influence. Darwin's work seemed to have the power to produce change. And the reason for this, Engels suggested, was that by calling something natural, it is made to sound eternal, right and proper. Any attempt, therefore, to suggest that the struggle for existence, war and competition were not the only means to secure human improvement could, following certain readings of Darwin's theory, be cast aside as going against the natural order of things. Such was the power of nature that once something was labelled natural, it could often seem that the right to challenge it had been extinguished. One way of representing this process of naturalization and depoliticization is to imagine a circuit of cultural and scientific production and reproduction (Figure 2.4).

   In short, Darwin's ideas about nature were, at least in part, shaped by the social world within which he worked. And, once published, the theory that Darwin produced was powerful enough to help lend further privilege to the section of society which Darwin represented (in particular the growing ranks of industrialists of the mid-nineteenth century, a new class that wanted to challenge the aristocracy at the same time as maintaining a sense of ordered change). The radical evolutionists had had their thunder removed by Darwin's theory. The latter could be read, and indeed was read, as giving support to those social evolutionists who had been writing prior to the publication of Darwin's theory and had been using Malthus to argue against welfare for the poor and needy. These Malthusian theorists of scarcity now had an authoritative scientific study with which they could boost their claims. Famine, war, and the elimination of certain races could now be considered 'natural'. While all of this was far from the intentions of Charles Darwin, his writings were useful ammunition for later generations of politicians and a number of social scientists who preached racial superiority, segregation, eugenics and even genocide (Livingstone, 1992).

Conclusion: From dependency to co-production

Thinking back to the three possible geographies of nature that I sketched in Chapter 1, it is now possible to see how these can be applied to this account of evolution. The first, independence, would seem to be undermined by our reading of evolution which was and is made sense of and acted upon in ways that are implicated within social worlds. As in Chapter 1, I have started to suggest that thinking of nature and society as separate or separable entities is hard to do in practice, and may well be futile. Nature's spaces are not straightforwardly independent of the societies with which they co-exist. A better spatial imagery than an island of natural facts untouched by people will be needed.

The second possibility was dependency, or the contention that nature is but a figment of human imagination and/or human social relations, something crafted in order to have particular effects. Nature is made to adhere to social orders and in being so conveys a moral authority upon those orders (see Daston, 2004, for a collection of essays on this theme). This is a wonderful ruse, or trick to pull. A double bind where nature comes to have the charac-teristics of the dominant way of ordering affairs, whose dominance is then propped up by an incontrovertible nature. The ruse is simple and effective, although not that effective, because we have just demonstrated what is going on. We have just performed something like a critique (an ideology critique). In doing so we have suggested something like this: nature is socially con-structed but is not widely recognized as such. It is thought to exist outside of politics (and therefore somehow is truthful, unbiased, the way things really are) but can be shown to be in part at least produced by politics and productive of politics.

Yet, there is a big difference between suggesting that nature and society are somehow connected, or relate together in one way or another, and saying that nature is dependent on society. For one thing, nature ends up looking very mal-leable, passive even. For another, nature turns out to be whatever society or the dominant versions of society, want it to be. Which is another way of saying that this is a form of idealism. The world is shaped to fit those meanings that are conferred upon it by powerful human beings. Indeed, in both possibilities we have considered, nature as independent and nature as dependent, nature ends up being a rather mute state of affairs. In the former it ends up being a solid rock of mute things that are entirely determinate and can therefore define what is real from what is imagined. In this version, nature is one single truth, and in being so it has to behave as a rather dull matter, doing the expected. In the latter (nature as dependent), nature has no say in what happens to it – it is shaped by people, or by the more powerful people, in their own image. Sometimes this might be a deliberate ploy, most of the time it will emerge as an outcome of the myriad of matters, languages, ways of thinking that are available to any group or society at a particular time and in a particular place (with all their connections to other places and times being just as important). But in either case, all the action is with the humans and their meaning systems. There's nothing very exciting about nature. It's putty in their/our hands. Surely there must be something more interesting about a topic, a place, that has excited so many commentators and agitated so many people? Surely there must be more to nature's spaces than a large vat into which we can put all the stuff that doesn't answer back, that does what it's told, that stays calm in all weathers? The next two chapters carry this forward, and start to build geographies of nature that can enliven human *and* nonhuman worlds.

Background reading

If you are not used to the idea of nature as ideology, then useful introductions include Hinchliffe and Woodward (2000), Neil Evernden's (1992) *The Social Creation of Nature*, and Noel Castree's (2005) *Nature*.

Further reading

For more on Darwin and the political construction of evolution, see David Harvey's (1996) *Justice, Nature and the Geography of Difference*. Adrian Desmond and James Moore's (1991) work on Darwin's life remains one of the best sources for understanding his life, work and times. Janet Browne's work is also immensely rewarding in this respect (Browne 2003a; 2003b). Robert M. Young's (1985) *Darwin's Metaphor: Nature's place in Victorian Culture* is a wonderful history, and Haraway's (1992) commentary on Young's work a very useful adjunct.

The edited volume *Uncommon Ground* (Cronon, 1996b) contains a wide variety of essays which unpack the ideology of nature as it is mobilized in areas of science, landscape, food, consumption, and so on. A criticism might be, however, that many of the essays tend to cede power and liveliness to social and cultural worlds, and leave everything else looking rather pale in comparison.

For a philosophical account of the problem of either nature independent or nature dependent, or in her terms 'nature endorsing' and 'nature sceptical' perspectives, see Kate Soper's *What is Nature?* (Soper, 1995).

Towards the co-production of nature and society

Independence and dependence, these first two possibilities might seem to be direct opposites. The first suggesting nature is out there, unsullied by the noise of humans as they meddle with their political machinations and other minor acts. The second suggesting nature is but a constructed place within society, and is thereby a result of the main business of economics, social relations, and so on. That's how things are often presented – a choice between a crude natural realism (nature independent) and a crude social idealism (nature dependent).

Yet another tack that started to be taken at the end of the last chapter was to suggest that these two possibilities, dependence and independence, are actually two sides of the same coin. And they amount to more or less the same thing. Both tend to produce a mute, inanimate nature, one that either provides a fixed bedrock to truth or one that simply gives in to human volition. Either way, nature is not up to much. This is a complex point, but the argument is that these two approaches to nature, which seem to be at opposite ends of the campus, are in fact the same. Those who emphasize the social shaping of nature (nature as dependent) do little other than tell us that nature is represented according to the societies in which those doing the representations live. But, in telling us this, in saying that our knowledge of nature is determined by social matters, one of two things can happen. The first, actually quite rare, possibility is that nature is dematerialized and becomes simply produced by social humans. We could call this an extreme idealism, and the point here is that nature vanishes into thin air (or into a vacuum). More commonly something else happens. In mounting an ideology critique, in telling readers that Darwin was influenced more by the practices of pigeon breeding societies south of the Thames than he was by his work on the Galapagos islands (Hinchliffe, 2000b; Secord, 1981), to say nothing of the undue influence of Malthus's texts and a concern to limit the radical potential of evolutionary thought, the implication is that the knowledge produced was *shaped by society*. Whether or not you agree that this is inevitable, the point is that there is an assumption here that the knowledge of nature is being polluted, or watered down, by social and/or political matters. And, the inference is often that nature itself remains unmoved by all of this huff and puff. We are simply talking about the shaping of the lens onto nature, rather than the shaping of natures. In other words, there are numerous ways in which the world is read that are socially sensible, but the world itself carries on regardless.

Which means that an ideology critique can also be a form of crude realism, in that the real world is unaffected by human scrapings at its surface. Or, at the very least, once you get beyond these surface scratches, nature stays the same.

Bruno Latour puts this rather well for our purposes:

> We are all familiar with the ravages of social Darwinism, which borrowed its metaphors from politics, projected them onto nature itself, and then reimported them into politics in order to add the seal of an irrefragable natural order to the domination of the wealthy. [This and similar examples] of ties between conceptions of nature and conceptions of politics are so numerous that we can claim, with good reason, that every epistemological question is also unmistakably a political question.
>
> And yet ... [w]hen one speaks ... 'about human representations of nature', about their changes, about the material, economic, and political conditions that explain them, one is implying, 'quite obviously', that nature itself, during this time, has not changed a bit. (Latour, 2004b: 33)

In this chapter we will expand on this killing of nature (for to suggest something doesn't change is to kill it off). We will look in more detail at some of its causes and consequences. Once we have understood this killing spree we can, in subsequent chapters, start to find resources for doing something a little more exciting than this dead end nature of social construction versus natural realism (or rather their alliance in making nature unhistorical, unlively, unchanging, aspatial and uninteresting). The aim of all of this is to make nature more interesting for all parties.

Divisions – their causes and consequences

After the first genetically modified primate was unveiled in 2001, one of the scientists said 'We're at an extraordinary moment in the history of humans' (Williams, 2001: 5). It was unremarkable, perhaps, that nothing was said on how extraordinary this might be for the history of primates. And nothing was said about how this might alter *geographies* of people, primates and a whole host of other humans and nonhumans. You couldn't get a more anthropocentric statement, rooted in a peculiar set of divisions and understandings of a primacy of human time, and a passivity of nature's spaces. If such a statement normally seems to pass without comment in our lives, then that is because there is a powerful set of divisions on which such a privilege rests.

To start to understand these common divisions we will use the example of a set of diseases called transmissible spongiform encephalopathies (TSEs). One of the most common of these is scrapie, which affects sheep. The best known is Bovine Spongiform Encephalopathy (BSE), or so-called mad cow disease. Both have relatively recently been associated with infective proteins. For many, the notion of infective protein might pass without notice. For

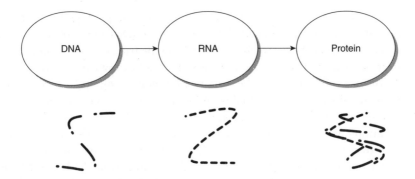

Figure 3.1   The central dogma of molecular biology

some, if not the majority of, biologists used to working with what has become known as the central dogma of molecular biology, the notion that proteins may have agency, that they can perform infection, is something of a heresy (Keyes, 1999a; 1999b). In brief, the central dogma of molecular biology, as described by James Watson, states that DNA makes RNA makes protein. For Watson's co-worker, Francis Crick, the dogma states that sequential information can pass from nucleic acid to nucleic acid, from nucleic acid to protein, but not from protein to nucleic acid, or from protein to protein. In short, the arrows in Figure 3.1 cannot be reversed, and the process has to start again for new protein to be produced. This implies that proteins are made, they do not make. The term central dogma sounds pejorative, but it was in fact a term coined by Watson and Crick to suggest a robust basis for molecular biological explanation.

Before the still controversial notion of infective proteins emerged from laboratories, it was generally assumed that an infectious disease required a self-replicating organism, one bearing nucleic acids (like a bacterium or virus), or the code for life, in order to replicate. But if proteins can cause infection, then here we have mute matter being lively. In telling some of this story of animated matter, our question will be one that Latour (1999) asked of Pasteur's mid-nineteenth century isolation of a yeast that explained the fermentation of lactic acid. Namely, did the microbes, or in our case these clever proteins, pre-exist their laboratory life? Did they come before or after the words to describe them, or do we need a more subtle narrative, and by implication something more usable than the geographies of nature that inhabit idealism (nature comes after the word) or realism (nature pre-exists the word)?

Scrapie: a sociable history

Scrapie is a disease of sheep that has been endemic in some places for over two hundred years. It is classified as a transmissible spongiform encephalopathy, a classification it shares with bovine spongiform encephalopathy (BSE), a disease of cattle, and Creutzfeldt Jakob Disease (CJD) which affects people. These classifications have largely been made on the basis of visible symptoms, as, over the course of the last century, very little was known about the causes of the disease (other than, as the classification name suggests, it can pass or be transmitted from an individual to individuals of the same species).

Scrapie has existed in laboratories, in mice and in various laboratory preparations as well as on farms. Using laboratory animals infected with scrapie, over the course of the twentieth century, people have looked in vain for a bacterium, for hereditary susceptibilities, and, what had seemed the most likely of agents, a scrapie virus (Keyes, 1999a; 1999b). Looking for a virus involves a whole suite of stories, practices, technologies, animals and people. It involves, or performs, what we can refer to as an *assemblage*, an active combination of technologies, ways of proceeding, their arrangements and their ongoing, unfolding nature (see Law, 2004a: 41). Assemblage is a term that is taken from the philosophy of Deleuze and Guattari (1988). It is similar in some respects to terms like discourse or episteme, both of which are associated with Michel Foucault (1970). However, where episteme or discourse tends to figure a way of doing things that can be rather too coherent, structured and set limits on what is possible, assemblage is potentially a little looser. An assemblage is in the process of forming and in so being, it is open to more possibilities. It is changing in ways that are not already determined by those things, people and stories that make it up. In that sense, assemblage has no truck with explanations of events that rely on either pole of the culture/nature divide.

One end result of the remarkably successful assemblage that is known as virology is in many cases a virus. As infective tissues are isolated and progressively refined, or made purer, the outcome is often a discrete, replicating bundle of nucleic acid that can be moved or can move from one host organism to another, often retaining an ability to produce disease in the new host. For those interested in nature–culture, a question at this point is, does this end result (the virus) exist before virology? If nature is a preexisting, independent and bounded space, then the answer would have to be yes. In the language of exploration, viruses would be said to be *discovered*. But if the virus is solely the product of the assemblage of virology, then our answer would be no. Viruses are *invented* as a means to explain disease. Yet another, perhaps more satisfactory, answer would be that discovery and invention are generally means of allocating special properties to either one of nature or culture. And therefore another way of looking at this is to suggest that as scientists started to interact with viruses, not only did science change, viruses also changed. This might seem rather odd. Surely viruses are viruses and we either see them (with our sophisticated techniques in the twentieth and twenty-first centuries (Figure 3.2)) or we

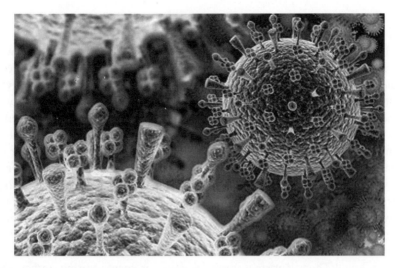

Figure 3.2 An image of the avian flu virus – discovered, invented, or neither?

don't? Given this common-sense approach, why bother to even say that viruses were not simply discovered?

One place to start is to say that we no longer talk about other peoples being discovered. As colonial histories of past and present have demonstrated again and again, discovery rarely leaves the discoverers and the discovered unaltered by the experience (Driver, 2001; Gregory, 2004). Human identities, we're reminded, are relational, produced through and with others, and not simply the product of inner make-up (Pile and Thrift, 1995). This compliment can also be paid to non-humans. When we do so, the poles of nature and culture become inadequate to our understanding of the history of science and the historicity of microbes (Latour, 1999: 146). The either/or of discovery/ invention is jettisoned in favour of something less divisive, more confused. This is difficult to express in abstract language, so in order to understand the co-activity of microbes and scientific assemblages, we can carry on with the scrapie story.

In the scrapie example, things did not quite work out as well as virologists might have hoped. As materials with higher and higher concentrations of infectivity were isolated, a process that would, if things followed the expected course of events, lead to 'pure' virus, something was missing. Contrary to expectations, the blueprint for replication, nucleic acid, was seemingly absent from the laboratory-produced infective material. Without this, it was difficult to imagine (within the assemblage of virology) how the microbe could manage to carry the instructions necessary to produce the effects of an encephalopathy from one organism to another. There are two issues that I want to pull out of this story of a badly behaved experimental object.

First, the object that materialized was not as expected. The scientists were *not in control* of their experimental system. But, just as this led some to continue the search for the elusive virus or some other nucleic acid-carrying entity, others started to listen, feel and work out ways of interacting with these materials in ways that unsettled the virological assemblage. Indeed, it was the ability to listen to the vagueness of what the historian of science Hans-Jorg Rheinberger calls the epistemic thing (Rheinberger, 1997), or the putative object of inquiry, which was a condition of possibility for new knowledge. (In putting together epistemic with things, Rheinberger is in some ways following the same course as Deleuze and Guattari, and Law, in suggesting that adding together epistemic and technologies makes for a more open structure, an assemblage. Rheinberger also adds that science in this light becomes more akin to a question-generating machine, rather than an answering machine.) The scientists weren't falsifying or corroborating, rather, they were tinkering, problem solving, interacting with apparatus and materialities of all kinds, writing and thinking. Another way of saying this is that the experiments were objective – but not in the sense of discovering, without the hindrance of ideology or culture, the real make-up of nature. Rather, those scientists that managed the experiments so that a different materiality could be engaged were being objective in the sense that they allowed the infectious materials to object to the stories that were being told about them (Latour, 2004a). The culture, or better the assemblage, of virology was important, but not simply because it framed the experiments, it was important because it too was changing as a result of a whole series of actions and interactions performed by the scientists, the apparatus, and not least the infective materials. Virology (culture) would never be the same again. Culture was being reshaped by nature, perhaps. These poles of explanation are in any case too purified for our purposes.

This brings us to the second point that it should not be presumed that because the research was objective (allowing the experimental objects to object to the stories that were being told about them) that we can appeal to a single, natural object in order to explain the experimental trajectory (or can say that culture follows nature). While it is true to say that the experimental objects were not *any*thing – they were not hapless, shapeless materials, or even fashion victims waiting for the virologist to give them any old form – I don't want to give the impression that these infective materials were timeless, natural objects (already formed and the same wherever or whenever they went). In other words, while it would be wrong to say that this infective material was culturally invented, it would also be wrong to say that they were *a-social*. Particularly if the term social is understood as an ability to associate (Latour, 2004a). Indeed, if we understand natures as *sociable*, then their importance to the experimental system, and as it turns out their less welcome sociability across species boundaries and through an industrial-agricultural system, becomes easier to imagine. Just as virology would never be the same again – so it is with these non-viral infective microbes. *Their* history was also changing as they interacted with feed manufacturers, cows, cats, people, politicians, ministries, zoos, burger franchises, trading partnerships, and so on (Hinchliffe, 2001). This busy socializing was in some way or another (and the histories and geographies are far from clear) linked to the events that have become known as mad cow disease (see Chapter 6). So it is also worth pointing out that these histories had already been written in an experiment that was bigger than the ones going on in the laboratories (Latour, 2004a). The collective experiment called, variously, the agricultural-industrial food production system, had already helped to provide the conditions of possibility for different histories and geographies for these microbes – and there can be little doubt that they returned the favour in the sense that the food system too might never be the same again. Finally, as these new infective agents took shape in laboratories and in our food system, Nature would never be the same again. It was now possible to talk about information transfer (disease transmission) in the absence of nucleic acid (DNA or RNA). In short, established theories of biological agency and the central dogma were starting to change (Keyes, 1999b).

Culture and Nature changed as the assemblage was performed (by humans, machines, proteins, and many others besides). It would be wrong to give all the credit to inventive, creative people or to immovable objects – both are in process, in assemblage (it's more informative to read assemblage as a verb in this sense). The scrapie story is useful to start us thinking about the utility of dividing the world into cultures and nature. It has also started to suggest that there are other ways of conceptualizing the relationships between people and things. But before we pursue some of these it is worth saying some more about the divisions, for this is not a simple trick that is played. Divisions work in a number of ways, sometimes to give Culture all the cards, sometimes to yield truth and reality to nature. It is useful therefore to construct a brief typology of divisions.

Division 1 Celebrating human autonomy

Immanuel Kant (1742–1804), in his *Groundwork of the Metaphysic of Morals,* said that only men are 'free with regard to all laws of nature, obeying only those laws which they make themselves' (Kant, 1948: 97). He called this unique ability 'autonomy' – literally self-law. Everyone and everything that didn't have the freedom to rule themselves, the will to make their own world, was heteronomous – literally 'other-law' – governed, if you like, from without.

Table 3.1   DIVISION 1 FREEDOM FROM NATURE

Autonomy	Heteronomy
Freedom	Governed by outside forces
Man	Woman, animal
Mind	Body
Reason	Force
North European	Mediterranean and non-European
Culture	Nature
Moderns	Pre-moderns

This neat world produces some neat spaces with clean boundaries. The main boundary lies between the human mind and the rest of the world, although other boundaries take shape too. Not everyone gets to be a man. There are forces outside the mind, bodily, emotional or nonhuman, that make this a select club. All of which means there is a human and physical geography, a body geography, a sexual geography and, as rationality follows a distance decay model as we leave Kant's home in Konigsburg, there's a neat geopolitics too. Table 3.1 shows Division 1, the freedom from nature. The left-hand column contains items thought to have transcended nature. The right-hand column shows items caught in its strictures.

TABLE 3.2   DIVISION 2 FREE NATURE

Free	Inhibited
Rural	Urban
Natural	Civil
Pre-moderns	Moderns

Division 2 Lamenting human autonomy

Around the same time (and I'm taking all kinds of liberties here), those who travelled a bit further afield than Kant (who famously never left Konigsburg), started to come up with another moral geography. Let's call them 'nostalgic Romantics' and say that they started to lament the lack of freedom that the cultivated men of cities expressed compared to their free comrades tilling the fields or basking on Mediterranean shores. For these men, then, another table might be appropriate (Table 3.2). In Table 3.2, the left-hand column contains items that were natural and free from social structure, while the right- and one contains items that were governed by society.

Now, in terms of the kinds of division being made, Divisions 1 and 2 look almost the same. Freedom on one side, chains of some kind (be they nature or civilization) on the other. A main difference is simply that nature switches from one team to the other. It is on the side of restrictive structure one moment and then on the side of freedom the next. Another way of saying this is that Division 1 celebrates the freedom *from* nature, while Division 2 celebrates the freedom *of* nature. No doubt the two divisions work well together, at once desirable and repulsive, the basis then for all manner of sexual and psychological work, and for some neat political manoeuvres. And it is this ability to render nature as both subservient to but at times transcendent of culture that forms a key trick in the politics of nature (see Division 3). Meanwhile, both divisions associate nature with pre-modernity. In both cases they are making an argument that, in this linear history, humans, or at least those who call themselves modern, are *'after nature'* – one division celebrates this departure, the other laments it.

Division 3 The politics of Nature

Another division owes more to Kant's theoretical philosophy than to his practical philosophy, and also owes more to those who came after Kant and who adopted his division of autonomy and heteronomy, which, as Wolfe (2003a) has demonstrated, in his exposure of multiple humanisms in western philosophies, is indeed a rich tradition. While universal truth and morality could only be derived from human freedom in Kant's practical philosophy, Kant was clear that such freedoms had some pretty secure physical limitations. Indeed, account had to be taken of the object world, for failure to do so would lead only to illusion. In other words, the sensed world had a part to play in the production of knowledge, but it was only a walk-on part. Latour has chronicled this nicely:

> Kant had invented a form of constructivism in which the mind-in-the-vat built everything by itself but not entirely without constraints: what it learned from itself had to be universal and could be elicited only by some experiential contact with a reality out there, a reality reduced to its barest minimum, but there nonetheless. For Kant there was still something that revolved around the crippled despot, a green planet around this pathetic sun. (Latour, 1999: 6)

So, according to Latour, in this new Copernican revolution, whereby the world is made to revolve around the human mind, nature is far from being eradicated, but it is numbed. For Latour, nature was taken hostage, and blunted, and brought into the service of epistemology as a means to adjudicate on the truth of statements. It became the matter of fact, the incontrovertible, the point at which all the protagonists should, if they follow the rules, agree that the time for debate is over. Any time there is a fear or risk of proliferation of viewpoints, of a multiplication of options, configurations or debates, the dead weight of nature is summoned. Latour has chronicled this ruse of the modern constitution for a number of years (see, for example, Latour, 1993; 1999; 2004b). In this political constitution people were given free will, but just in case they exercised it too freely, reality was made into a hard and fast, solid and incontrovertible check. Nature in this form is most useful as a means to secure (for a time at least) agreement and to justify one view over others. In short, the world external to the mind becomes a bedrock which grounds truth (the truth that is universal, timeless and applies everywhere). In this grounding function it can be relied upon to adjudicate on matters of controversy, by supplying matters of fact (Latour, 2004b).

It may be that Latour's reading of Kant is rather too neat here, or at least it may be that he is overstating the case. For example, Jane Bennett offers a rather less dismissive reading. She compares Kant's nature to that of the Renaissance physician and alchemist Paracelsus (1493–1541), working some three hundred years earlier. Paracelsus was, Bennett tells us, a kind of Christian animist, combining 'the idea that plants and animals are powerful agents with the idea of a heavenly Creator' (Bennett, 2001: 35). The details are not our concern, but what is important for Bennett is that Kant maintains some of this liveliness, even while, under the influence of Newton, he is keen to remove the divine purpose from matter. Bennett goes on:

> Nature for Kant is not altogether different from nature for Paracelsus. For neither of them is nature dead matter, devoid of marvelous powers, for neither is the natural world disenchanted. Kant's nature speaks, though more haltingly and cryptically than Paracelsus. But it utters enough to assure us that, or at least give us enough hope that, the world is a coherent order. (ibid.: 44–5)

Even if the blame for the death of nature can rarely be pinned on one philosopher (and Descartes is another who often comes in for rough treatment with respect to nature's fall from grace), it is nevertheless pertinent to suggest, as Latour does, that the moderns succeeded in a double move of relegating nature to matters of fact at the same time as making sure that humans obeyed the resultant universal laws of nature. Examples of this kind of division of the world into warring social matters and coherent, ordered and passive nature abound. Some are pernicious while others are well-meaning and possibly benign. From blaming death and destruction on a so-called natural event (the discourse of natural disasters – see Box 3.1) to justifying universal human rights on the basis that we all share one nature, matters of fact are rolled out to absolve some and sometimes all of us from the hard work of politics. Yet, just as in the case of understanding the diseases of sheep and cattle, it becomes less and less useful to rely on such divisions. Matters are sociable rather than natural or social. And being sociable they can change.

Box 3.1   Natural disasters?

How convenient it was for Bush senior and his son, George W. Bush, when they were presidents of the USA, to label Hurricane Andrew (1992) and Hurricane Katrina (2005) natural disasters. In both cases of course it was the socially vulnerable who suffered, and in the latter case the wafer-thin social contract in the United States that contributed massively to the disaster. Hurricanes are of course examples of the liveliness of the world, and are born from spatial multiplicity (and it is possibly not unimportant that the composition of the atmosphere is changing rapidly as a result of human actions, to which US contributions in the form of greenhouse gas emissions are famous). Just as the causes are neither purely natural nor purely social, the effects are also co-productions. Prior to the breaching of the levees, which protected the below-sea level city of New Orleans, George W. Bush's administration had cut the budget for maintaining flood defences by 50 per cent. Similarly, the political and personnel resources at the Federal Emergency Management Agency (FEMA) had been reduced to the point of making it unable to respond effectively. There can be little doubt that these changes to resource allocation had effects. There are many possible stories that can be told about such a disaster. To label something a natural disaster is to start to sort those stories and to absolve politics and politicians from taking matter(s) seriously. It is to excuse the failure to address all the contributions to suffering.

Conclusion

Independence and dependence, discovery and invention, the natural and the social – these are ways of explaining events, like diseases, scientific claims and disasters, that in turn suggest a bi-polar world, one where culture and nature exist in separate and separable realms. We have looked at some of the rationales for these divisions – human autonomy versus nonhuman heteronomy, cultural possibility versus natural necessity. The attractions of a partial human exceptionalism coupled with a natural brake (so, as Latour puts it, humans, or the mob, the demos, don't get too carried away) are easy to see. It's a ruse that has served democracy, or mob rule, rather well. But there has been a cost. In this chapter, I have started to suggest, through the example of scrapie disease, that purifications are not only difficult to achieve, they also start to have deleterious effects on the ways in which responses to disease and other events are made. They make matter mute, and in doing so shut off all manner of possibilities, from human wonder at the world to experimenting with ways of living that are of greater benefit to a range of earthly inhabitants. These points are expanded upon in Chapter 6 which looks in some more detail at the BSE or mad cow crisis, and highlights the deleterious work of dividing nature and society. Prior to this, however, we need to complete the tour through nature's spaces and consider in more detail the third possibility that nature and society are co-productions. Our question becomes, what kinds of geographies of nature can exist in this co-produced world?

Background reading

A basic discussion of the divisions between the natural and the social and the political work to which these are put, especially in the case of not so natural disasters, can be found in Hinchliffe (2000a).

Further reading

Bruno Latour's *Politics of Nature* (2004b) provides an excellent account of the relationship between the natural and the political, something that is also developed in *Pandora's Hope* (Latour, 1999). The latter also includes a fascinating commentary on Latour's earlier work on the historicity of microbes. For a slightly different account of Kant and the possibilities he affords and does not afford for a lively nature, see Bennett (2001).

Hybrid natures

For some time now the term 'the matter of nature' has held sway over geographical imaginations and practical work. Fitzsimmons' (1989) timely call to nature has had human geographers running for the hills (and rivers, genomes, animals, and so on). What Fitzsimmons anticipated was that the matter of nature was no easy matter at all. Indeed, and above all, nature does not and cannot easily be located, described or used. It is not self-evident. Matters of nature are not matters of fact, or indisputable realities (Latour, 2004b) that can somehow ground geography, or secure political agreement. Just as the 'what' of nature remains something of an open question, it follows too that there is no obvious answer to the 'where' of nature. As we have seen, it's not simply out there, untouched by human hands. Nor is it in here, formed in human minds. A third possibility is that it is something that is enacted, or co-produced as I roughly termed it earlier. This chapter and Chapter 5 investigate this possibility further.

The overarching aim is to answer the following question: what kinds of geographies of nature can be said either to exist, or can be enacted, now that the *old* divides of nature and society have been abandoned? What other spatialities can there be for nature? We have, in the previous chapter, followed Latour in suggesting that natures, like cultures, have histories. Now we need to start to explore whether natures, like cultures, have geographies. Another way of posing the question is to ask whether, as Annemarie Mol has put it, things (like bodies and natures) do not simply have a contested history but also a complex present, 'a present in which their identities are fragile and may differ between sites' (Mol, 2002: 43). The question, I will argue, is a crucial one with numerous practical implications for the development of environmental policy, conservation politics and food safety (no doubt among others). Some of these issues will be the subject of later chapters. Here I am more concerned to start to unpack what is meant by the notion that nature is practised, is a co-production and is multiple. How can we understand this? I look at two resources for thinking about this in this chapter: interaction and hybridity. In Chapter 5, I discuss a third possibility: difference.

Interactions

If nature and society are not necessarily marked by hard and fast divisions, then maybe we can think of them as interacting. On first blush, this seems to make sense. The natural world interacts with the human world as soon as it is enunciated, encountered, used, valued or whatever else humans do to engage it. One could read, for example, the account I gave in Chapter 3 of the development of the laboratory-based scrapie-disease assemblage as an example of interactionism. There I asked the question, 'Did prions exist prior to the lab work?' In other words, did the scientists invent them out of thin air or simply discover them? Neither answer satisfied. Instead of imagining an all-consuming (or all-producing) social construction of a passive nature, or a world of self-evident objects that remain unchanged and wait only to be revealed by humans, an alternative account was provided. We followed Latour in tracing the histories of things, of their relations with, among others, human beings. To butcher the story a little more, prions, or something like them, did exist prior to their articulation in laboratories in the USA and Scotland. But they were not the same things that existed after those experiments. They were affected by their experience, just as they were affected by changes to the ways in which animal feed was manufactured in Britain and by the architecture of the ruminant feed and the specified bovine offal bans after the onset of BSE (see Chapter 6). Prions, like microbes and other things (Latour, 1999: 150), have histories. They change and are changed as they interact with other parts of the world.

This sounds reasonable. But before anyone supposes the problem is solved, that interactions of nature and society hold the key to understanding the world, there are a number of recurring issues that present themselves:

First of all, while apportioning a role to interaction in the shaping of nature cultures, we only do so by assuming that somewhere, buried deep in time or in the structures of these forms, there are pre-existing natural and cultural properties (to be clear, this is in no way the kind of story that Latour would tell, far from it, but it is one that can appear from time to time in the literature). In other words, while the forms we observe, like prions, are neither pure nature nor pure culture, they can nevertheless be storied as the result of interactions between pre-existing constituents, one of which is nature independent. So the problem is not so much solved as put into the background (only to re-emerge on closer inspection). The tendency is to ship all the creative work back into primary interactions between nature and culture. Forms (be they children, microbes or farming patterns) then become the predictable outcomes of those interactions. That is they are *determined* by the things interacting. Thus a form of determinism (see Box 4.1) holds sway. As Wilson (1996) demonstrates, in psychology, such interactionist claims only momentarily overcome those interminable debates concerning 'nature or nurture' and end up attempting to quantify the extent to which traits can be ascribed to genes (nature) or to environment (culture). And even if this sounds, at the very least, to be symmetrical in terms of distributing causal powers or agency across the human-made and natural-made worlds, then such an impression is soon lost when we note that the two terms, nature and culture, have not shifted one nanometre. The result is that we have two pre-existing, mutually exclusive realms that can and do interact, in deterministically simple or even complex ways, to produce new forms. Even then, primacy tends to be given to one side or the other of this re-stated divide. For Wilson, writing from psychology, it is nature that tends to be considered as the least ephemeral of the prime movers. As she puts it, 'nature is the foundational bedrock to which culture brings a series of secondary inscriptions' (Wilson, 1996: 58). Elsewhere, culture is the more intransigent actor, producing nature after its own image (Castree, 2004).

Box 4.1 Determinism

Determinism is a form of philosophical reasoning which holds that every event, including matters as diverse as cognition and hurricanes, can be explained by a causal chain of other events and processes. There is no randomness, no magical input, no mysterious term that can be used to account for an event. In geographical thought the most common expression of such reasoning has been environmental determinism, the suggestion that human activities are controlled by the physical environment (Glacken, 1967) itself playing into and fuelling a form of naturalized North European supremacism (Livingstone, 1992). In studies of science and technology, a good deal of work has been done to unseat technological determinism – the belief that, for example, social, economic and political forms could be explained by the fixed properties of technologies. In both cases, the form of argument is similar in that variations in material form and function are used to explain variations in social and cultural activity. And, the former variations can be explained with reference to first principles, or, in other words, to nature. Another type of determinism is historical determinism which holds that irrespective of actions taken, events will unfold along a pre-set trajectory. The latter is sometimes linked to structural materialism and is associated with Marxism. However, it is important to note that not all forms of Marxism are determinist. Indeed, rather than demonstrating the highly structured linear narratives of historical determinism, Marxism is at its most productive when open and multiform flows render new possibilities for political action practicable. On the possibilities for these more open or aleatory Marxisms, see Bennett (2004).

A second and related problem is that, in being deterministically produced, these 'interactive' forms are rendered as passive entities. Children, landforms, prions, whatever, can end up looking like nothing but manifestations of two active realms which interact in more or less complex ways. All the action is passed backwards to the formative processes that are either nature or culture. These entities are then products of history and geography and have no hand in the making of those histories or geographies. All of this is tantamount to suggesting that things really don't have history or spatiality at all. For if things are but manifestations of already settled matters (nature and culture), then their form is already describable prior to the event of their gestation. Which is the same as saying that really nothing has happened. Things are simply doing what they are told, or what has already been determined for them. Things are just surface manifestations of already settled matters. At another resolution, nothing has changed.

To pursue this point, the trouble is that in this account of the world, nothing new could actually ever happen. It is an argument that suggests only that matter unfolds in time, along already prescribed trajectories. There is, as Doreen Massey has argued, no space in this kind of account, and no possibility for other trajectories or indeed, co-existing and coeval trajectories (Massey, 2005), or even the possibility that trajectories will swerve (Bennett, 2001). As Massey also insists, without space there can be no time either. For if time is made through the creation of novelty, then for there to be time, things must be open, exposed to other things. So in this closed world, that simply performs to a script, nothing really happens and there is little point trying to effect change. The trouble, then, with interactionism is that there's no time-space for difference. There is precious little action in interaction if it is conceived as a relation between pure forms.

For all its promise, interactionism leaves nature as an independent entity, deterministically producing interactive forms that seem to have no real say in matters. For these reasons, I want to leave interactionism as a non-starter. It's a weak attempt to leave the shores of independent nature and seems only to return us on the next tide. The second strategy considered here marks a more successful departure from this world of dualism and determinism.

Hybrids

The term hybrid refers to the progeny that result from the interbreeding of unlike kinds. In Latin the term *hibrida* was used for the offspring of a tame sow and a wild boar. Fear of hybrids is one reason used against the reintroduction of wild boar to the British Isles, even though hybridization hasn't occurred in that part of south-east England where a few wild boars roam (see MacDonald et al., 2000). The baggage that the term carries, both in this gendered meeting of passivity and activity, of already existing domesticity and wild life, is compounded by the sometime racist mobilization of the term to refer to impurity.

Despite this questionable baggage, the term hybrid has more recently been rendered in different directions. It has been used to figure a relational geography and social science wherein the mixtures and configurations of machines, animals, states, organizations, ecologies, politics are continually made up of all manner of elements, which themselves are nothing if not hybrid forms. The last point is important. It helps to jettison the notion that forms are a result of the combination of already existing and completely self-contained kinds, though this risk still remains, and some authors prefer terms like 'crossings' in order to avoid the image of hybridity as 'static entities coming together to form a compound' (Bennett, 2001: 31). The important thing to note is that the aim here is to avoid another form of determinism, something that beset interactionism. Rather, in the relational geographies that are figured by the mobilization of hybrids and crossings, there is nothing outside the mix (including a pre-formed Nature), and, importantly, parties are re-configured as they relate or engage with one another. So this is different to interactionism in that we are not talking here of pure forms that mix to produce something that is reducible to constituent bits.

Likewise, this is different to dialectics. For even though many of the authors who are working with metaphors of hybridity share a good deal with dialectical thought (there is still plenty to recommend in David Harvey's (1993; 1996) dialectical programme, in part, because he works with many of the philosophers who have inspired hybridity theories), there is nevertheless a tendency in dialectics to systematize, to render relations as contradictions and to eventually pose nature and culture as pure ontological categories that no one can reconcile (Latour, 1993: 57). The versions of dialectical thinking that have taken hold in the social sciences have been beset by dualisms, not least the modern settlement of nature versus culture but also the reduction of any difference to conflict and thereby to contradiction. Even the attempts to develop green versions of Marxism (Benton, 1989; Dickens, 1992; Grundmann, 1991) seem beleaguered, in the last instance, with these problems.

The metaphor of hybridity allows for something different, it allows for change in all parties as they relate to one another. And it allows for novelty to be produced. Novelty that is not reducible to component parts. Indeed, parties do not simply interact to produce a new (impure) form. Rather, in relating, the parties and the product must change too (this is the key to most versions of relational thinking). Nothing remains unaltered in the event of relating. So, as Donna Haraway has put it in her account of the hybrid naturecultures of dog–human relations, the 'relation is the smallest unit of analysis' (Haraway, 2003: 4). Similarly, Sarah Whatmore (2002) mobilizes a relational ontology in order, among other things, to reconfigure wild(er)ness as a more spatially complex set of relations. Building on Cronon's (1996a) engagement with the wilderness tradition in North American environmentalism (see Chapter 1), Whatmore seeks to abandon the 'syntax of distance

Figure 4.1   Wild boar or tame sows – pure nature and culture or always already hybrids?

and proximity; inside and outside; then and now' (2002: 11), that has inhabited exotic environmentalist divisions, in order to offer an alternative, topological, folded sense of wilds. So, 'the notion of wildlife being fleshed out here is a relational achievement spun between people and animals, plants and soils, documents and devices in heterogeneous social networks which are performed in and through multiple places and fluid ecologies' (ibid.: 14). And, it should be emphazised, those people, animals, soils and others start to look very different when regarded as already hybrid forms in process. In Figure 4.1 which shows a boar-cross the actual 'status' of animals is often contested. There is no agreed way to make absolute decisions of purity or hybridity, although disputed schemes exist for differentiations based on number of chromosomes. Even these are contested. The image above shows an individual whose snout is too short to be pure boar for breeding purposes and is probably the result of a domestic cross. The wild boars of south-east England turn out to be the progeny of escaped zoological animals. No pure form there, just as it would be ridiculous to argue that domestic pigs are purely of the house of (agri)culture.

So wherever we go, be it the land grazed by elephants close to the Okavango delta, or to Central London where pigeons scavenge for food (to take two places that might seem poles apart in the divided landscapes sketched in earlier chapters), matters of all kinds (places, people, representational devices, and so on) are folded into the mix. Which is to say that things, including so-called wild and domestic settings (be they national parks, animals, body tissues, or legal frameworks), are not in themselves wild or domestic. They are hybrid forms, more or less durable bodies made up of similarly hybrid and impermanent relations. Things are, to use another commonly used term, configured, or drawn together, in order to become more or less stable forms. There are no pre-existing essences, only relations.

Hybrids and the term configuration take us some way to overcoming the dualisms of our divisive Kantian world (see Chapter 3). An ontology of division is replaced then by an ontology of configuration. The ontology allows Callon and Law (1995), for example, to suggest that we understand agency, that term so often associated with the autonomous human beings of the Kantian settlement, as 'effects generated in configurations of different materials' (ibid.: 502). This is a distributed sense of agency, one quite different to conventional attributions.

> Attributions which localize agency as singularity – usually singularity in the form of human bodies. Attributions which endow one part of a configuration with the status of prime mover. Attributions which efface the other entities and relations in the *collectif*, or consign these to a supporting and infrastructural role. (ibid.: 502–3)

There's a politics here, then, that contests conventional attributions of agency and suggests an experiment in the fermenting of new collectives, which are less sure of themselves, less convinced that it is people who call the shots or that things are already determined in nature.

The bias that is built into the divisions that I sketched in Chapter 3 results in attributions of agency being almost always placed upon human beings, sometimes on machines and less frequently on animals (or specifically, animals that to some extent resemble human beings in their neurological apparatus or at least in their outward form). Sometimes, as we have also seen, a thing called nature can be storied to take centre stage. But the issue is not that agency can be extended now and again, it is that, for Callon and Law at least, such attributions are effects, sometimes useful effects, but we should not necessarily confuse processes of attribution with processes and relations that can make things happen. Things happen through hybrid *collectifs* and not as a result of pure thoughts. Tellingly though, agency tends only to be attributed to point locations. 'So fields, diversities, processes, or areas – whatever the metaphor might be – agency isn't attributed to these' (ibid.: 497). Our ways of making sense, then, tend to figure particular species and spaces of agency, but such spaces cannot be the only story to tell when we're interested in what happens and what might happen now. In other words, we need different spatial imaginations in order to take account of, work with, intervene in, hybrid worlds.

Box 4.2   Social topology for hybridity – regions and networks

What spaces for a hybrid world? If you imagine that nature and culture are independent matters (see Chapter 1), then, as we have noted, you envisage them as islands separated by a gulf. To think like this is to think in regions and volumes (and only in this way). In this version of the world, regions are bounded, internally pure, and take up a certain amount of space. As Mol and Law put, 'Space is exclusive. Neat divisions, no overlap. Here or there, each place is located at one side of a boundary. It is thus that an "inside" and an "outside" are created. What is similar is close. What is different, is elsewhere' (1994: 647). Regions may bump into other things (like snooker balls bump into each other), change course, but essentially they stay the same.

   If, on the other hand, you imagine things as hybrids, then regions and vol- umes can only be part of the story. It becomes necessary to understand how things are connected to and disconnected from other things. How, in other words, things are related. One way of doing this is to say that things form in and through *networks*. This is a term derived from semiotics (the study of how meaning is built), and used to understand how realities are built. Following the work of those who developed material semiotics (Akrich and Latour, 1992), networks are used to understand how realities are built through matters as diverse as machines and gestures, wires and glances, bodies and codes, numbers and poems. Building meaning or building realities is not a matter for pure things. Just as conventional semiotics tells us that words mean in rela- tion to other words, so too realities are built from the networking of all man- ner of matters. In the case of material semiotics, though, it is not meaning that is the issue. Rather, the elements in the material semiotic network allow each other to work or to do things (Mol and Law, 1994: 649). And each element is configured as it enters relations with other elements of the network. In this way an elephant in a national park (Thompson, 2002), a water pump in Zimbabwe (de Laet and Mol, 2000), a prion in animal feed, a vet on a dairy farm, a wolf in a pack – they don't work or do things all by themselves. All are configured through their network relations with others, and change as they move or are moved around. This is one way that Actor Network Theory (ANT) tends to develop accounts. Matters are in relation, they are not pure forms. Networks fold together what might have formerly been understood as distant things. Elephants and population calculations start to look closer together in the network view than they would do in the regional view. Prions and govern- ment offices start to be uttered in the same breath. No longer belonging to either nature or culture, networks allow us to see things as cuturenatures. Crucially, things may exist in more than one spatial form. In one telling,

(Continued)

a prion or elephant may be thought of and enacted as a region or as a container, in others it may more usefully be thought of and enacted as an effect of a network. Annemarie Mol and John Law (1994) drew upon and reconfigured the mathematical term 'topology' to articulate this possibility that realities are made of more than one spatial type.

Is everything hybrid?

All of this seems reasonable and does move us beyond the divisions of society and nature in potentially useful ways. However, let me discuss two problems that are highlighted in the literature. First, there is a fear of flattening everything out, producing a world where everything is related to everything else, with no tools available for differentiating matters of importance, political or otherwise. As we will see, this is a well-recognized problem, and it leads into a second issue. *If* we are to reject this undifferentiated field and in its place look to a world where things are more differentiated, then how is this differentiation to be done? How to make and mark a difference? I will now take each of these problems in turn.

Even though the spatialities have seemingly become more complex, so much so that configurations and hybrid forms travel the world and are no longer confined to bounded times and places, we can, if we're not careful, end up with a seamless world in which it becomes almost impossible to envisage or enact anything at all. To be sure, relational thinking does a lot of work, but what it can also do is replace an ontology of division with one of force – a Nietszchean world where everything is in *principle* related to everything else (Hetherington and Lee, 2000). In this there is sometimes no space left for the unmapped, for those that do not lend their force to the emerging network in any unproblematic manner. Nick Lee and Steve Brown highlighted this problem of Actor Network Theory, in particular, but relational thinking more generally, by suggesting that ANT had no 'other'. They read ANT as a version of liberal thinking, extending agency to all. In a simple sense, following this logic, everything is exhausted by the network metaphor, and with everything related to everything else, it was totalizing (Lee and Brown, 1994). This was not just a problem of semiotics, part of the problem was the route through which actor-network theory and feminist science studies derived their semiotics – from Greimas's structuralist analytic that tended towards an all-encompassing systemic generation of meaning through relation (Haraway, 1993; Hinchliffe, 2003; Lenoir, 1994).

Does it matter that there's no outside to networks? Massey has argued that without space, there is no multiplicity, and without multiplicity, nothing can happen. The same can be said for 'things' and networks. The philosopher Graham Harman argues this point very nicely. He suggests that if we reduce 'the being of objects to their relational situation' (Harman, 2002: 229), if we privilege the network of negotiations between things (something he accuses Latour of doing), then we end up with a single world, where nothing really happens. 'If the cosmos is truly one, devoid of any *genuine* specific regions, then nothing would conflict with anything else, and the universe would resemble a placid lake' (Harman, 2002: 292, original emphasis). His argument is that things are never simply made up of their current set of relations, there is always something in reserve, something that withdraws. In other words, there has to be space for difference. 'If an entity always holds something in reserve beyond any of its relations, then it must exist somewhere else. And since this surplus or reserve is what it is, quite apart from whatever might stumble into it, it is actual rather than potential' (ibid.: 230). Now, as Harman is at pains to show, this is not to suggest that there is some core or essence, some old-fashioned unchanging substance to things. It is more that there is an outside to networks. So, against any 'severe form of holism, we need to re-establish the firewalls that protect every entity from its neighbours. To do this without relapsing into a conservative version of substances may prove to be one of the great challenges ... over the next several decades' (ibid.: 256–7).

So how has hybrid thinking dealt with an outside? Contra Harman, Latour has an outside to the business of associations. To be sure, all matters are capable of forming associations but many do not, remaining outside the collective which is continuously in formation. These outsiders can, Latour argues, put the collective at risk and thereby contribute to the process of change (Latour, 2004b). Even so, Latour is adamant that prior to the formation of a collective there are no pre-conditions that dictate membership of a collective. In other words, flatness is a principle that should be upheld. Arguing against the tendency of social science to assume a pre-existing realm of three dimensional images: 'spheres, pyramids, monuments, systems, organisms, organizations' (Latour, 2005: 172), Latour advises that we flatten things out in order to see how dimensions are generated. While this is attractive as a methodology (and one that I readily attempt to use elsewhere in the book) and while the subtleties of Latour's work are clearly not to be underestimated, it is also indicative perhaps of a tendency to see the world mainly in terms of its associations and dissociations, of insiders and outsiders. Meanwhile it is the homogeneity of the former which underplays the rich variety of partial connections, loose affiliations, deferrals, allegiances and differences that also inhabit and make the world.

Elsewhere there have been numerous attempts both to see the outside of networks and to develop understandings of social processes that are also alive to difference (Callon and Law, 1995; 2005; Mol, 2002). These authors are working towards understanding difference in ways that I take up in more detail in Chapter 5 and subsequent chapters. For now, I want to trace a risk in some of the work that attempts to devise an outside to networks. The danger is that in desiring an other we can inadvertently return to some fairly conventional divisions. We can, as Latour would put it, move too quickly in adding volume to the flat lands of relational thinking (2005: 171). There are many examples of this tendency that could be used but I will focus on one of the more interesting and demanding, that of a more than human phenomenology. I will use Whatmore's (2002) *Hybrid Geographies*, not because it *ends up* making such divisions, in fact, it skilfully avoids them, but because along the way the tendencies are plainly evident.

A more than human phenomenology?

Whatmore usefully employs the network metaphor of ANT to multiply the actors and actants and to diversify the spatial registers through which naturecultures are produced. This is effectively the first of two manoeuvres that are employed to take us from pristine (and passive) Nature to a more promiscuous wild-life (2002: 14). The topological approach (see Box 4.2 on regions and networks) helps to 'unsettle the contours of these exteriorisations of the wild by situating them within the diverse currents and flows through which multi-sited wildlife networks are configured' (ibid.: 14). The breeding programmes and protocols for zoological animals, with their international co-ordinates, data banks, movements of tissues and re-introductions, attest to this multi-spatial, and temporal, hybrid geography. The second manoeuvre is the one that provides a departure from some earlier relational geographies and starts to reintroduce difference into these currents and flows:

> The second manoeuvre involves *animating the creatures* mobilised in these networks *as active subjects* in the geographies they help to fashion. Their constitutive vitality is acknowledged not in terms of unitary biological essences but as a confluence of libidinal and contextual forces. Here, the multi-sensual business of becoming, say, antelope or wolf and the inscription of these bodily habits in the categorical and practical orderings of human societies are interwoven in the seamless performance of wild-life. (ibid.: 14–15, emphasis added)

This re-animating project is most obviously taken up in a chapter that seeks to work with elephant geographies, and thereby on animals that are described as 'saturated with being' (2002: 36). From what I have said so far it should be clear that I wholeheartedly agree with the attempt to partially differentiate fields in just the kind of way that Whatmore sets out. However, I think there are a number of issues which are not quite resolved in the strategy adopted here and which risk returning us to an unhelpful ontology of division.

Most tellingly there is a potential distinction being made here between lived experience (which, contra Descartes, higher animals can be said to participate within) and non-living matter. So for instance, it is non-human animals that are saturated with being, that exhibit subjectivity and that can act as genuinely disruptive figures, as agents who can unsettle the networks. These figures are used to generate an explicit distinction between an emerging animal geography (Wolch and Emel, 1998), alive with albeit hybridized differentiation, and those social studies of technology and science, where attributions of technical agency are more easily told as relational effects, though Latour's (1996) anthropomorphic treatment of a non-existent light transit system tends to unsettle the distinction (see also Philo (2005) on this). Phenomenology, itself a broad term used to figure a philosophical movement aimed at describing the structures of experience as they present themselves to consciousness, can tend to install an experiencing being as the foundation of action, even if that action has lost a Cartesian cogito. Indeed, while Descartes is vanquished successfully, there is a residual Kantian settlement in this work which tends to divide the world itself against the phenomenal experience of that world. The risk is that a privileged relation to relations is reproduced that seems only to result in re-inscribing human beings as the most experiencing of all, and inevitably leaves most of the world back in that neverland of dull and mute matter. In drawing explicitly on phenomenology (even a more than human version), there remains the danger that the poles of nature and culture remain firmly in place, even when they are now re-labelled as producing nothing more than an insurmountable tension – see Latour (1993: 58). Harman captures the problem well when he says that Heidegger, that most 'environmental', but as Harman (2002) and Giorgio Agamben (2002) also demonstrate, in the end, anthropo-centric and anthropo-logic, of phenomenological thinkers, 'seems to think that human *use* of objects is what gives them ontological depth, frees them from their servitude as mere slabs of present at hand physical matter' (Harmen, 2002 : 16), and he

> wrongly casts Dasein [roughly speaking the 'being there' of human subjects] in philosophy's starring role, while preserving the unfortunate belief that the world itself is made up of sheer physical objects; neutral slabs of material accidentally shuffled around or coloured by human viewpoints, stable substances volatilized only by an external force. (ibid.: 19)

For Harman, a phenomenology of things requires a backgrounding of the knowing subject. If we want to follow phenomenology's, and its instigator Husserl's, lead in getting to the things themselves, then we need to unsettle the central role given to humans, and to a few carefully chosen fellow species who can also experience, in bestowing meaning on mute objects. As Whatmore suggests later in her book, as it becomes more Deleuzian in orientation, differentiation is more complex than something that experiencing bodies can effect on a world of inexperienced and mute matter. We need other ways of understanding and working with animal spaces, technical spaces and many others besides, that do not return us to Kantian subjects and to such stark differences in kind (see Chapter 9).

Politically and ethically, elevating animals from the flat lands of relational thinking may be strategically useful. But I want to push things a little further. For there is a risk that we are inadvertently returning to *a priori* divisions between the living and the non-living, between legitimate subjects and dead objects. In relocating agency at a point location of animal, we retain a set of privileges to the relations that informed the initial analysis (and by default, perhaps, human privilege is maintained as they experience most of all). The question is, can this be pushed a little further for I'm not convinced that animal politics or a more general 'ethics and politics committed to the flourishing of significant otherness' (Haraway, 2003: 3) are best served, even strategically, by retaining what is effectively a language of human privilege that is selectively extended to a few fellow travellers.

The broader issue in this chapter has been to explore geographies of nature. The hybrid metaphor has enabled us to understand spaces of co-production a little more clearly. But on its own it is too undifferentiating, and tends to obliterate natures rather than offering means to understand their production and productivities. The point here has also been to register discomfort with any return to an experiencing being as the means by which we can differentiate the world. In order to develop an alternative, I will need to explore an ontology that builds on hybridity, but which tries even harder to refuse the re-settlement of the landscape with sub-Kantian subjects.

Conclusion

In this chapter I have looked at two main ways of dealing with the co-productivity of natureculture. The first, interactionism, was rejected for a variety of reasons, not the least of which was the tendency towards dualism, determinism and thereby unconditional closure. Much more promise was found in hybrid metaphors, as long, that is, as they aren't read in ways which presuppose a pre-existing pure form. However, as this and the next chapter suggest, hybridity cannot be equivalent to undifferentiated mixing – for such mixing can only produce a placid surface whereupon nothing can happen. In other words, hybridization is too general a term, we need to think about other ways in which things meet up, miss one another, connect, disconnect, accommodate one another, form collectives, and so on. We may also need to think through how it is that things withhold from relations, and maintain a distance. So how to understand differentiation becomes the key. I have rejected any easy recourse to experience or being as the means by which we can differentiate the world. The next chapter elaborates on other possibilities.

Background reading

Sarah Whatmore's (2002) *Hybrid Geographies* develops many of the themes in this chapter and the next. The opening chapters are more in line with the way I have represented hybridity in this chapter, the final chapter has more in common with the way I take this forward in Chapter 5.

Further reading

Donna Haraway's (1991a) *Simians, Cyborgs, and Women: The Reinvention of Nature* is a fascinating performance of hybridization, presenting a political call to mobilize the resources that are attendant to this way of thinking and acting in the world.

Bruno Latour's (1993) *We Have Never Been Modern* remains one of the best examples of hybrid thinking, with its clear justifications of this approach over more conventional social theoretical devices (including modernism and postmodernism).

Geographies of nature and difference

I now want to argue that there are geographies *of* and therefore spaces *for* natures. Not pure spaces in the manner that I have sought to criticize in the chapters thus far. But neither spaces where there are only mixtures, or where everything becomes related to everything else, replacing, as Hetherington and Lee (2000) point out, an ontology of division with a relational ontology of force, and therefore another version of the same (the same because as we have seen in both these versions, 100 per cent purity and 100 per cent mix, nothing new can ever actually happen). Rather, there are practical, empirical and, to be sure, political spaces of variability. Spaces for and of animals, bodies, rocks, office blocks, manifestos, and symphonies – in short, spaces of and for difference.

As we have seen in previous chapters, there is a political need to abolish an incontestable, transcendent Nature. However, as the previous chapter suggested, some kind of element of surprise, some form of otherness, is vital if we are to avoid an undifferentiated field or a hyphenated, list-like, social science. Following Massey (2005), I have argued that without multiplicity, it becomes impossible to produce change. One tempting solution is to return to a relational sense of experiential being, but I have hinted that this risks a number of problematic divisions. In this chapter I want to draw on alternative experiments in becoming where the divisions are less readily drawn back to Kantian settlements.

In order to experiment with the difference of nature that I am after, I need to go over some material that initially at least draws us back to the previous chapters, and to Kant's distinction between autonomy and heteronomy. I want to make two sets of arguments. First, human beings have never been autonomous, and second, nonhumans have never been heteronomous. This allows me to suggest that rather than looking for stark *differences in kind*, we can instead work with *differences in degree*. We are all of us, human and nonhuman, living and non-living, different by degree. Once the 'of a degree' argument has been made, I want to start to outline how differentiation is not only possible but also inevitable. In the account I draw upon two overlapping traditions for thinking through difference before returning, in the second half of the chapter to an account which most readily spatializes, and thereby allows me to start to re-build, nature's geography.

Degrees and differences

First, 'we' have never been autonomous (or at least, autonomy is a process that involves far more than a self-knowing human subject). Here there is perhaps a remarkable convergence of philosophy, science studies, and bio-physical sciences, which moves us away from the constitutive figure of the human, 'or more properly speaking, toward an exposure of the human's own impossibility' (Wolfe, 2003b: xi). Humans act and speak to think, there are demonstrable half-second delays between initiating acts and the decisions to start them. (In Chapter 1, I mentioned that you don't run away from a dangerous situation because you are frightened – to wait until you were frightened would be to lose precious moments – you turn to run and then experience the frightening thoughts.) There are intentions in action (Gray, 2002; Massumi, 1996; Thrift, 2000). Minds are enacted in highly differentiated worlds. In other words, minds are part and parcel of the world, not control centres that somehow rise above it and dictate to it. Humans are as much hetero as auto, as much governed by 'outside' as they are 'inside'.

The re-working of Bergson by Deleuze and a host of others (see Watson, 1998, for a review) highlights the complex movements of matter that tie together humans and nonhumans, mind and matter, memory and experience, the virtual and the actual – so much so that any differentiation of consciousness, ideas, representations and matter can only be differences of degree and not of kind. Consciousness then is but a particular kind of movement in this material continuum. It, like all other matters and matterings, exists in what Deleuze terms a plane of immanence. This term is not equivalent to a fixed nature, rather it is perhaps best thought of as the open conditions for differentiation. So Deleuze's project, along with a number of other anti-juridical philosophers, notably Spinoza and Foucault (Gatens, 1996: 164), is in part to refuse a dualism of, on the one hand, a plane of nature or immanence (the given) and, on the other, a transcendent plane which socializes the first (in other words, it's not stuff on one side and consciousness on the other, but the stuff of consciousness).

> There is no dualism between the two planes of transcendent organization and immanent consistence ... [w]e do not therefore speak of a dualism between two kinds of 'things' but of a multiplicity of dimensions, of lines and directions in the heart of an assemblage. (Deleuze and Parnet, 1987: 132–3; see also Gatens, 1996)

The complexities of this work are manifold, but one consequence that can be emphasized is a rejection of final causes, or non-caused causes. It is a rejection, as Gatens (1996) puts it, and by way of example, of any attempt to explain gender by sex (attributing cause to Nature), or, conversely, of accounting for sex through social powers and the autonomous choice of gender. It is, in other words, a rejection of any account of the world which posits material necessity against a voluntarist idealism. As I have argued throughout, this choice of Nature or Culture is not one that I want to have to make, can make or indeed is one that can deliver in the fraught politics of nature. The plane of consistency or immanence is then a device for thinking and acting in ways that refuse these grand transcendences of Mind, Human being, Nature, God, Society or whatever else. But, to be clear, the plane of immanence is not equivalent to a placid lake surface. Everything is far from being of a piece.

If humans are governed as much from without as within, the reverse can be said of nonhumans, providing a second refutation of Kant's divisions. Indeed, it is now reasonably widely accepted that biological systems exhibit autonomy. The work of biologists Humberto Maturana and Francisco Varela has been particularly instrumental in demonstrating that organisms are continually self-producing (Maturana and Varela, 1992). They use the term autopoietic, or self-organizing, to suggest that organisms (so-called individuals as well as other collectives) are to an extent closed systems. While organisms are open to their environment, they are nevertheless self-referential. As Wolfe summarizes, 'biological systems are self-developing forms that creatively reproduce themselves by embodying the processes of adaptive changes that allow the organism to maintain its autonomy or 'operational closure' (2003a: 36). The autonomy that Maturana and Varela pose is, to be clear, very different to Kant's free agents. For them, autonomy and its operational closure are neither self-identical nor exogenous but a becoming that is about constant returns and circulations. Autonomy is a precarious achievement, and involves more than a self-contained agent.

To summarize, this convergence of understandings of becoming is deeply wary of hard and fast divisions. It is anti-Cartesian monism, and so allows for no separation of mind and body. It is connectionist, arguing that all is matter, only allowing differences in degree between perception of matter and matter itself. It is anti-representational, insisting that there is no occult mind using representations to mediate world and perception. In place of this, there is a world of complications, foldings and involution (a term I return to in the latter parts of this chapter) (Watson, 1998). In addition to its anti-Cartesian stance, denying that a material world exists 'outside', and a mysterious immaterial world of representations exists for the cogito, there is also a mistrust and denial of Kantian settlements. Rather than seeing a world in itself, on the one hand, and a phenomenal experience of that world, on the other, the work that I am drawing on here sees no originary distinction, no pre-existing division between organisms and environments, sentient beings and life worlds. Rather, differences *are made in practice*, in the process of making a world. Thus, we get to the importance of the virtual and the actual to Deleuze, of the world in its totality (the virtual) and that which is actualized (the actual). Both of these are material, the former offering possibilities for (rather than determining) the latter (again Watson, 1998, provides a lucid account of these and other components of the new non-dualist vitalism).

This is difficult territory, but what I take from it is that first of all there are no absolutely distinct vantage points from which the world can be viewed – in other words, there are no inevitable borders between human or sub-human experience and a world of things. Everything here is potentially lively. Second, and before this starts to look like everything is connected to everything else such that all matters can be reduced to their total relational situation, there are differences in the making.

Difference and semiotics

Another way of conceptualizing difference is to follow a more textual route, one more often associated with writers such as Jacques Derrida than with Deleuze, Latour and others. To introduce this work here might seem surprising given that difference in this sense is sometimes associated with a descent into a form of textualism that has little or nothing to offer a project concerned with the movements and crossings of matters. But the outcome can be quite similar and both traditions work in a semiotic register even if the initial emphases are rather different. In a nutshell, and to recapitulate, analysts have taken the following from semiotics. There are no fully present and locatable elements in a system. That is, identities are produced through relations, themselves made up of traces of every other element (Wilson, 1996). For the feminist neurologist/philosopher of science, Elizabeth Wilson, every element is always marked by the trace of every other element. Impurity is key. It is just this form of analysis that can allow us to respond to Nature 'not with repudiation ... or with a lurch to the cultural' but with natural overdeterminism that operates in excess of the limits of presence, location and stasis (ibid.: 59). What this means is that we should certainly refute the arguments of conservatives who would use a timeless nature to pronounce on their versions of sexuality, belligerence and markets (for example) as the only truth of the matter. But, the denunciation should not imagine itself as relying on purely cultural histories of these terms. It should develop arguments which emphasize the indeterminacies and vagaries of all matters, so-called natural as well as so-called social. It is the will to retain the *specificity* of nature at the same time as not allowing 'this specificity to coagulate into a sovereign autonomy that owes no debt to another' (ibid.: 60) that is key. This is a semiotics then that is, or can be, less overarching than its more structural versions. It's a looser weave, allowing for specificity, allowing for difference but without suggesting that those differences are pre-set or for all time.

Likewise, and rather like the Deleuzian account of difference sketched above, Wilson is insistent that we disallow any fully present inscriptive power to apply itself on to inert matter. In other words, there is no point at which we can infer an origin. The play of difference, or *différance*, a term Derrida derived to figure the always ongoing processes of differentiation and deferral, is crucial here and invites a particular form of stance or intervention in nature politics. Before I elaborate, let me use Wilson's characterization of *différance*:

> We must remind ourselves again and again that we are not situated at a cross-roads, we are not confronted by a choice. The problematic here is not one of either environment or neurology [culture or nature in the terms we are using here], either inside or outside, but rather the interminable, seemingly intolerable, play between them. The play is not the interaction of two seemingly discrete terms (nature plus culture); it is not the secondary manipulation of already existing entities. This play is not subject to resolution or quantification (how much play? with what fragments? in what order?). Rather this play, which Derrida also calls différance, is the condition of all 'neurological' and 'cultural' difference ... Différance does not come before neurological differences as their cause (thus the name origin no longer suits it) – it is that which allows the production of these differences and which remains in excess of that production. For this reason, différance cannot become present; it cannot become the object of scientific inquiry. As a non-originary origin, *différance* thwarts the desire to reveal, in biology or in culture or in the interaction of biology and culture, a final and incontestable foundation. (Wilson, 1996: 61)

So what do these kinds of claim allow us to say, and, more importantly, what do they allow us to do? Both accounts of difference share a number of features. First of all, they help to remind us that we are seeking something different to natural or cultural causes. Or better, we're no longer looking for a fixed property of nature, or of culture, to explain the world. Nothing is fixed, even mountains move (Massey, 2005) so much so that there is no nature to go back to, to call up as witness. There's no tracing back to first instances, to original causes, to natural-born or cultural-made things. For nature and culture alter all the time and all over the place. They don't combine just to produce culture–nature. When they combine, they are no longer what they were beforehand, they affect one another. So, as suggested in Chapter 3, prions affect nature (in the form of a threat to some biological orthodoxies) and they certainly affect what has been until now called culture (British beef is tainted with new associations). Second of all, difference and *différance* suggest that that we are not here talking of the flatlands of a hyphenated sociology, a land where everything can be related to everything else and becomes, therefore, an undifferentiated force field. There are differences, but, importantly, they are not reducible to any of the old categories (and this is where some of the piety of phenomenology may be found wanting, see Chapter 4). In other words, from Latour, Deleuze, Derrida and the other authors we have briefly engaged with, we can take a thread forward that we should be vigilant in attempting to avoid recourse to those grand transcendences, those a-historical matters (be they mind, Nature or whatever) that seem to provide first causes, primary properties, origin stories. For all these authors, nature has history, it changes. For some of them, this marks the end of the story for nature. If it is no longer a timeless matter of fact, then why use the term at all? The short answer is that natures have geographies as well as histories. And in doing so, in being spatial as well as temporal, we may well want to keep the term going.

Nature's geography

So there is more to the world than mixture. Difference might be the term to use. If it is, then we might want to make a distinction between two ways of thinking this difference. As Boundas (1996) and Massey (2005) tell it, 'there is an important distinction between *discrete* difference/multiplicity ... and *continuous* difference/multiplicity ... The former is divided up, a dimension of separation; the latter is a continuum, a multiplicity of fusion' (Massey, 2005: 21). The distinction comes from Deleuze and Bergson (as we have seen, their argument is for difference by degree, for a continuum, with no pre-existing, authentic, unchanging entities, or magical starting points). There is then differentiation; stuff happens, new things are created (in ways that are not determined by starting points). The continuum is an important device for refusing arbitrary starting points in any account of what happens. But the term needs qualification, otherwise it can sound as though we are back to an undifferentiated world of everything being related to everything else. In this section I want to start to put some detail on this multiplicity that has no origin but at the same times is not a fused, seamless world.

What must be emphasized at this juncture is that difference is not something visited on the scene by human (or another experiencing being's) perception or orderings of a world. Things happen even if no one knows anything about them. In other words, the continuum is not equivalent to an undifferentiated mass that becomes differentiated simply because some creature sorts the world into forms with which it can then live. It is already differentiated and in a process of differentiation. So it is important here not to imagine that all difference is but a convenient foil, an effect, made by those subjects who are working to order the world for their own devices. In other words, difference is not an arbitrary ordering visited on the world, with the concomitant job of analysts and social scientists being to debunk myths of identity or differentiation, showing instead that in actual fact everything is interconnected. Such a knowing stance is characteristic of a certain kind of critical social theory that enjoys puncturing illusionary balloons like identity, nature, human, and so on. As Barnett (2005) astutely observes (writing about the ethical import of this style of thought), in this generic version of poststructural social theory, which for Barnett includes hybridity theory, any seemingly watertight dichotomy is critiqued for its arbitrariness. In this style of working, all identities, be they nation-states, national parks, human beings, society, culture, or nature can be shown to be the product of relations. They are built, it is argued, from partial exclusion, denial and repression. The critic then has to show that all these absent things are in fact present – or that they are potentially there waiting to disrupt the imagined unity. The result, as Barnett shows, can be a rather high-minded critique of any form of identity or entity formation.

So how to avoid such a critique? For Barnett, one answer lies in a more complex account of division. Rather than assume that people, in his case, are constituted through the exclusion, repression and denial of others, others that will come back to haunt and disrupt their seemingly coherent identity (making everything that was pure seem impure), he follows Lois McNay's (2000) suggestion that exclusion isn't the only response to difference. Accommodation and adaptation might be just as likely as denial. In other words, difference is not only produced through exclusions, through making pools of order in an already thoroughly mixed world. Rather, it is already in and of the world and is dealt with in a variety of ways that cannot simply be described in terms of exclusions. In order to emphasize this capacity for dealing with difference, McNay draws us back to a human agent 'endowed with the capabilities for independent reflection and action' (2000: 3, cited in Barnett, 2005: 8). But we don't need to reintroduce this romantic figure of the capable and reflexive human subject to take something important from this work.

Another way into this issue, returning us to the topologies introduced in Chapter 4, is that just as there are limits to regional or volumetric ways of figuring identity and difference, there are also limitations to thinking in terms only of networks or only in terms of a fairly structural form of semiotics. The crucial issue is that things, be they subjects or objects (it really doesn't matter here), don't simply exist on their own as a pure form, but nor do they exist by virtue of excluding others. It is not simply that A is made by pushing out all that is not A. A exists through transformations which allow some degree of consistency through time-space and which involve all manner of relations with others which may not always be contradictory (A, not A). As Mol puts this, 'coexistence side by side, mutual inclusion, inclusion in tension, interference: the relations between objects enacted are complex. Ontology-in-practice comes with objects that do not so much cohere as assemble' (2002: 150).

This is not necessarily news to those in science and technology studies who have been aware for some time that things happen not simply by themselves (the regional view), nor simply by virtue of a set of stable and probably invisible relations (the network view). There may be other topologies. Mol and Law (1994), for example, suggest that in addition to the social topologies of regions and networks, things can also flow, maintaining their integrity while the world around them shifts. They take the example of anaemia in The Netherlands and in Zimbabwe. The treatment and diagnosis of anaemia occur in both settings. But that is not to say that they are the same. Indeed, they are rather different in terms of assemblages, techniques, symptoms and results. The laboratory-based diagnoses available almost as a matter of course in The Netherlands are at best intermittent in Zimbabwe. But there is anaemia in both countries and it is diagnosed and treated as such. Importantly, it isn't by virtue of a stable network that the medical practices of anaemia diagnosis and treatment can move between and within The Netherlands and Zimbabwe. Diagnoses move by means of more than a network. The practices of anaemia are not immutable mobiles (Latour, 1987) produced via a sturdy network. The network of machines, devices, numbers, practices and personnel who hold the nature of anaemia stable over what hitherto were regarded as long distances can sometimes work, but more often than not fails to produce a smooth diagnosis. Failure does not mean that anaemia, its diagnoses and treatments, disappears. Some things hold, despite losing some of their relations. Anaemia is diagnosed (in different ways with different results). The conclusion is that medical practice does not move out from a centre along a stable network – or if it does this is not necessarily the most robust or failsafe way of doing things. Rather, practices can be said to flow, to act something like viscous fluids that can reshape and regroup to adapt to different conditions.

> Isn't this a plausible way of talking about what happens to anaemia as the Dutch physician moves from the Netherlands to the new context of her clinic in Zimbabwe? For the 'anaemia' she learned about in medical school doesn't disappear as she steps on to the aircraft. Not at all. And neither does it shift, as it were like an ontological rupture. But the anaemia she finds in Africa isn't familiar either. So in the weeks after she arrives, she starts to absorb what's novel about it. She learns that there are no machines. Or that the figures they produce aren't accurate. But she finds that there are plenty of white eyelids, which she only read about in books before. And she discovers that there are pallid skins – though she can't 'see' that they are pallid. The local nurses are far better at this than she is. (Mol and Law, 1994: 658)

The point here is that 'the world doesn't collapse if some things suddenly fail to appear' (ibid.: 659). The anaemia assemblage can 'accommodate' and 'adapt' (ibid.: 662). It is receptive and thereby more durable than some of the networks that also result in anaemia being diagnosed. Another way of saying this is that the assemblage is obdurate, it lasts, despite being in many ways non-coherent. This is a point Law develops elsewhere (Law, 2006a), drawing on the work of Charles Perrow (1999), suggesting that loosely coupled systems are often more stable than those that are tightly coupled. Quickly put, things hang together precisely because they can exist apart. In the quote above, the work of accommodation is largely attributed to the human medic – but it should be emphasized that the medic does not do this on her own. Machines, numbers, colour charts and many others besides also adapt to and help to constitute different diagnostic conditions. Likewise, in a separate study, Marianne de Laet and Annemarie Mol trace another story of accommodation, this time very much from the direction of a nonhuman. They tell how a water pump travels and works in a variety of places by not entirely holding its shape (de Laet and Mol, 2000).

So how do we draw this back to spaces for nature? The new issue is this. Not only are we not going to reduce nature to society, we are not going to reduce *anything at all* to the network within which it resides. Material semiotics, hybridity, network – these terms are not of themselves enough to allow us to understand the way things work, don't work and the manner of reality that we want to call nature. At least, they are not sufficient if they are interpreted in ways that highlight stable structures or homogeneity. The point for Barnett and, in our terms more promising because they don't restrict accommodation and adaptability to people, for Mol and Law, is that there are continuities. Things can hold together or work through exclusion of others and through the construction of relatively stable networks, but this isn't the only way that realities are enacted. Things can remain consistent while undergoing significant changes, they can be motile as well as mobile (Munro, 1997).

In terms of natures, we can say two things. First, they are not already there, pre-existing. Rather, they are the outcome of actions (actions that may involve people, or may not). Second, natural matters, like any other things, are not reducible to their network relations. They can hold some of their shape, some part of their difference, even if parts of their supportive network crumble. But the consequence of this second issue is something we haven't yet considered. How can something, without an origin, nevertheless *not be* solely a product of its relations?

One answer is to suggest that things, including networks and relations, are never exhausted by their relations. In other words, things are only ever partially connected (Strathern, 1991; 1996). Or, to put things another way, there is always a portion of a thing, be it a dog, a table, an international programme to reduce anaemia, that withdraws from the networks within which it is partially formed. Something is held in reserve. This was Harman's (2002) point that we encountered in Chapter 4. He argued that for anything new to happen, matters must not be exhausted by their current arrangements, or networks within which they have been implicated. And to hold something in reserve, this thing must exist somewhere else (ibid.: 230). It is different. Which is not to say that it has a fixed identity – this is not to return us to a privilege of the regional, or the pure. Rather it is to say that things have histories (that much we know from Latour and others), but they also have geographies – they are partially connected, yes, but they are also partially withdrawn.

Another answer, which may be more promising in that it offers a more positive way of talking about materialities (rather than saying there is a remainder that is somehow not equivalent to substance), is the more empirical and practical approach of ontological politics. It is to say that things are made by more than one practice, a multiplicity that of necessity produces an instability and a potential for new configurations. Things are, as Mol has it, pushed and pulled into one shape or another (2002: viii), and in turn push and pull other matters. But rather than this being a simple ontology of force, the multiplicity of things throws up another dimension, one where things are not only shaped but also may have to cope with more than one shape. If they are practised in different ways, and if no one single practice achieves absolute dominance, then we are in a situation where any one thing is also multiple. They will need, in other words, to live with and by difference.

Box 5.1   Multiple things – the short case of insulin

Here is a fairly standard story of insulin. Insulin is a form of protein that many mammals produce in their pancreas. Once secreted, it becomes attached to receptors which are present on most cells of the body. The attached insulin then activates other receptors on the cell to absorb glucose (sugar) from the bloodstream. Without insulin, or without its proper attachment, this process is impeded and the patient develops a form of diabetes (there are two main types of diabetes, Type 1 resulting from a deficiency in the production of insulin, Type 2 from a failure of insulin to attach to cells). Type 1 diabetics and to a lesser extent Type 2 diabetics rely on insulin being introduced into the bloodstream in order to maintain safe blood sugar levels. For many years this insulin was extracted from animals, from cows and pigs, as a by-product of the meat industry. However, while this insulin works very well for most people, some develop allergic reactions. Their immune system recognizes that the insulin is from a 'foreign' body. A solution to this problem was to find a way of producing human insulin. All humans have the same form of the insulin protein, so an immune response would be less likely. In the 1980s, the development of recombinant DNA techniques allowed for the mass manufacture of this product. This involved the cloning of the human gene which codes for the insulin protein, and its subsequent insertion into bacterial DNA. The modofied bacteria are then encouraged through various means to continuously produce insulin, a process that is scaled up by pharmaceutical companies to mass-produce human insulin in large vats.

Insulin has a history, it has been made in lots of bodies for a very long time. It was first made in a laboratory in the 1920s by Banting and Best along with their dogs. It was made in slaughterhouses and by pharmaceutical companies in the mid-twentieth century as a by-product of meat production. And it was made in the 1980s in laboratories and then vats using transgenic

(Continued)

bacteria. In some versions of this history of a thing, insulin becomes less natural and more human-made over time. But that's only one telling. What if we think less about the origin of something and more about what it does? What if we think less about time and more about space (or better, what if we think about time and space together)? What we can start to say is not only that this potted history records changing involvements between people, pigs, bacteria and insulin, but it also necessarily involves a multiplication of insulin. So while insulin starts to become more and more like an entity that can circulate outside the body it also (1) becomes entangled with many more things; and (2) becomes more than one thing.

There are two elements to the story of insulin that are important for geographies of nature. First, the history of insulin is a story of becoming natural. The insulin produced in the vats of the pharmaceutical companies is said by some to bear 'no trace of its artificial origin in the laboratory' (Harman, 2002: 249). For Harman, it is not natural in the old sense of falling out of the sky with no history or social pedigree. It's not pure region. It has its networks of association. But, it is natural in 'its manner of reality' (ibid.: 249). It is able to move from the vats to human bodies and perform as if it was pancreatic insulin. Harman is partly right here, but the rhetoric of no trace underplays the complex realities and spatialities of entities. So we need a second point.

Second, the insulin not only becomes more and more thing-like, it does so through rather than despite its relations, and through rather than in spite of its multiplicity. Insulin wouldn't work on its own (take away the bacteria, the vats, the doctors, the needles, the current forms of capitalization and property rights enjoyed by pharmaceutical companies, and it would stop being available until such time that other forms of organization could be put in place). Moreover, injection is not a simple or stand-alone practice. Diabetics also need to engage with other practices including monitoring and measuring diet, taking exercise, learning to be affected by bodily sensations, maintaining needle hygiene, and so on (Mol, forthcoming). And, the use of extra somatic insulin is not without potential complications (while its laboratory trace is not necessarily evident, its injection may cause short- and long-term side effects such as hypoglycaemia and changes in the distribution of body fat). In these senses, pharmaceutical insulin is not a smooth matter of fact – it does, contra Harman, carry traces if we include these economies and complex bodies. Insulin has some rough edges. But, manufactured insulin also leaves the needle as a reality that can, in many senses, do as good a job as pancreatic insulin. It flows into the patient's blood. The network enacts an

(Continued)

(Continued)

entity that can travel, it can circulate (Latour, 1988). It can act at a distance. It is, as Latour would call it, an immutable mobile (Latour, 1987). It holds its shape as it moves along the network, from vat to bloodstream. But that can only be part of the story. To 'work' insulin needs to combine with protein receptor molecules at the cell surface. The combination then activates all manner of cellular reactions for a period of time until the combined molecule is absorbed into the cell and broken down. The point is that insulin is far from being inert. It is stable enough, but its action is also dependent on many other things being in place (if the receptor system doesn't work effectively, then the patient also suffers from a form of diabetes). And it is the combination, or mutual accommodation, of insulin and reception molecules, that makes insulin work in the body. While the pharmaceutical network creates and is created by an immutable mobile, a relatively stable form of insulin which, given the right temperatures and storage conditions, can circulate through hospitals and homes, the action of insulin is only effective once it combines with an accommodating body. Accommodation implies some form of adaptation, some molecular co-formation. In other words, insulin also flows. To an extent like the physician getting off the plane in Harare, it is part region, part network and part fluid. And to be all of these it can relate, withdraw, hold its own and adapt. To do all of these there must be multiple insulin, and insulin must be the fragile assemblage of multiple human and nonhuman practices. The remarkable thing about the insulin story is not then just the history of associations but the geographies of associations, accommodations, withdrawals, multiple practices, and so on that cohere enough to keep the molecule working.

Involving geographies of nature

In Mol and Law's terms, we might say that things work through inter-topological effects, through regions, networks and fluids (1994: 650). I have added to this topological analysis the point that, in order to work through such inter-topological effects, things must adapt and change in ways that may be partially irrespective of a network within which they co-exist. And in order to be able to do this, things cannot lend themselves fully to any single network. I have also rejected the argument that sometimes reserves this adaptation and accommodation to human beings alone, or even to those entities that are saturated with being. All things are made of a variety of relations, some of which will include the ability to withdraw or be partially connected (Strathern, 1991). In Massey's (2005) terms, there has to be internal and external multiplicity, and thereby there must be space.

One distinction I have started to make here is between metaphors of mixings, of relations and kinship, to one of partiality. It is at this point perhaps that the work of Deleuze and Guattari (1988) becomes most useful. In Chapter 3 we used the term 'assemblage' to refer to the active combination of technologies, ways of proceeding, their arrangements and their ongoing, unfolding character. In particular, it was the openness that this term implied that was most attractive in terms of understanding the creativity involved in laboratories. The scrapie protein was not, I noted, a product of nature or culture, but the outcome of an open framework which allowed for the emergence of difference. Rather like insulin, the scrapie and later prion proteins became more real as the assemblage grew. But they also became multiple, and thereby not reducible to a single network.

Another term that is useful here, the progressive meaning of which is derived from Deleuze and Guattari (1988: 238), is 'involution'. The term evokes the contagion of the world, and it is rather different to the standard notion of evolution or change by filiation, by hereditary. The logic of the latter is present in many versions of hybrid. Marking the imbrications of human and nonhuman, it runs the risk of turning us all into monsters, of producing a single identity, 'lessening the differentiation of the two' (Baker, 2000: 126; see also Whatmore, 2002: 161, for the difference here between Haraway and Latour's use of hybrid). The alternative mechanism of involution signals a becoming whereby species are folded together to form blocks of becoming. 'There is a block of becoming that snaps up the wasp and the orchid, but from which no wasp-orchid can ever descend' (Deleuze and Guattari, 1988: 238). It is evolution by symbiosis rather than filiation. So rather than transgressing identity with monsters, involution sweeps them away, to talk of alliances. What we get then is change through contagion – a contagion that is partial and not all-consuming. The alliances that create are not therefore inter-breedings. Rather, these geographies of nature are heterogeneous, they are of unlike kinds that participate. In some senses this is like hybrid, but the emphasis is rather different. Deleuze and Guattari spell out the distinction:

> Contagion and involution are like hybrids, which are in themselves sterile, born of a sexual union that will not reproduce itself, but which begins over again, every time, gaining that much more ground. Unnatural participations or nuptials are the true Nature spanning the kingdoms of nature. Propagation by epidemic, by contagion, has nothing to do with filiation by hereditary, even if the two themes intermingle and require each other ... The difference is that contagion, epidemic, involves terms that are entirely heterogeneous; for example, a human being, an animal, and a bacterium, a virus, a molecule, a microorganism. Or in the case of the truffle, a tree, a fly, and a pig. These

combinations are neither genetic nor structural; they are interkingdoms, unnatural participations. That is the only way Nature operates – against itself. (1988: 241–2)

The only way that Nature operates is against itself – another way of saying this would be to insist that for anything to happen, there is involution, contagion, multiplicity and space. And that rather than contagion and involution leading from different identities to the same, from the many to the one, differentiation generates difference. This apparently simple conclusion is a vital notion for nature's spaces.

This is of course an argument for nature, or for spaces for nature. It is not necessarily a description of how things are, or how things are currently turning out. Indeed, some of the evidence around us seems to contradict this wild philosophy of differentiation. Biodiversity seems under threat, landscapes are tending towards the uniform rather than the differentiated, everywhere increasingly seems to be characterized as the same rather than as different. Meanwhile, with a different set of spectacles, the world seems more exuberant than ever, productive of all manner of crossings. My point in raising this seemingly contradictory tale of two natures at the end of this argument regarding natures's spaces is that we clearly do not need general statements on the state of nature. There isn't necessarily more of it or less of it now than in the past. The point of this chapter and the ones that have preceded it is to suggest and emphasize the practical point that spaces for nature can exist. Their shape is now hopefully a little clearer than when I started. We have certainly become quite good at saying what they are not (spaces for nature are not independent, or singular, nor are they wholly dependent). The more positive argument is more difficult, but some progress has been made. Spaces for nature are formed through multiple practices, from partial connections and from involutions. As matters of practice, and multiple practices at that, they are empirical matters. So in Part II we turn to empirical cases, and at the same time expand on the implications of geographies of nature.

Conclusion

I started by saying that there are spaces for nature – but they are not obviously independent in the sense that I traced them in earlier chapters. So how to evoke spaces for nature was the question. In this chapter and in Chapter 4 three main possibilities have been reviewed, nature–society interactions, hybridizations and difference. They are all, to some extent, versions of relational thinking. We have ranged over material that is seemingly pulling in similar directions. But I have been careful not to allow us to be drawn back to those forms of human-centred thinking that allow human beings a privileged relation to the relations that make the world. All of us (bricks, books, proteins ...) form alliances, but we do so in ways that do not exhaust 'all' of us. In other words there is duplicity or multiplicity in relations (rather like the ways in which we sometimes accuse untrustworthy characters of duplicity). Which means that there are spaces for nature. That is, there are spaces to spring surprises.

That nature is spatial as well as temporal is a claim that we will want to explore further in subsequent chapters. Each of these has an empirical core. The reason being that investigating spaces for nature is an empirical-theoretical task. There are no universal spaces of nature, no equivalent to the independent state of nature that was thought to apply everywhere and for all time. My argument is that if there are spaces for nature, then they will be multiple. The possibilities are numerous. So the following chapters start to unpack some of these possibilities as they relate to particular cases.

Background reading

Much of the work in this chapter is underpinned by actor network theory, and particularly the attempt by practitioners to come to terms with difference and otherness. Nick Lee and Steve Brown's formative paper 'Otherness and the actor network: the undiscovered continent' (1994) is a very good place to get to grips with the issues. John Law's book *Aircraft Stories: Decentering the Object in Technoscience* (2002) brings ANT together with Deleuzian under-standings of difference. For the clearest treatement of multiplicity, see Annemarie Mol's (2002) ethnography of *The Body Multiple*.

Further reading

Sarah Whatmore's final chapter of *Hybrid Geographies* (2002) outlines some more detail of this Deleuzian understanding of hybridity. Doreen Massey's (2005) *For Space* is a rich resource for thinking seriously about how we spatialize social theory and how social theory and philosophy can underplay time-space.

Part II

How and Why Geographies of Nature Matter?

Part I started with a fenced-off woodland in Manchester. There was and is something of an ambivalence in terms of my response to that woodland. On the one hand, I hate to see it placed off limits, but on the other hand it is still there, much as it was when I was a child, partly as a result of the fence which has done its job of keeping some people out. Even so, perhaps we should take the fence down and open it up to people, hoping that this show of trust would produce a sense of responsibility, stewardship and care. This would result in the woodland being practised in different ways, with different management plans, people, animals and other matters becoming part of the pulling and shaping of the woodland.

The point here is not to use the arguments in Part I to make a council's decision for them, for this would always be a complex political process. Rather, it is to highlight that in thinking about the woodland, it is helpful to talk about its spaces rather than its space, its multiplicity rather than its singularity and the need to find ways of moving forward while recognizing that more than one woodland is being enacted. There are no magic wands which can solve the problem of the fenced woodland. There is no independent nature nor any social critique that can sanction the boundary or its removal. Rather, there are numerous woodland practices, not all of them with people at the centre, that need to be assembled carefully and in broad accordance with which practical ways forward need to be devised. These practices will be far from perfect, not agreeable to all and may turn out very different to how they were imagined, but that is probably the best we can do.

What is crucial here is the detail of the situations in which nature is made multiple. Part II therefore visits a number of case studies in order to expand on our understanding of geographies of nature and how and why they matter. My broad aim is not only to provide empirical examples that demonstrate and extend the discussions in Part I, it is also to start to offer some answers as to how to live together with multiple spaces for nature. In Chapter 6, I look at the relations between nature, science, policy and politics, taking up the case of BSE, already introduced in Part I's discussion of prions. It's a case study of what goes wrong when a singular independent nature is used to bypass multiplicity, and thereby science is used to avoid doing politics. Chapter 7 looks at the issue of biosecurity, highlighting the multiple ways in which diseases and geopolitical securities are practised and the effects this has on the kinds of policy we can expect to work. Chapter 8 compares the practice of conservation with its theory, focusing on urban natures. If natures are not wholly separate to the world in which they are practised, then how best to move conservation forward? Chapter 9 extends the question, 'how can we live together?' (Latour, 2004b) by looking at various possible relations between people and animals, and considers the possibility for learning from these interspecies relations as a way to generate a broader environmental ethic. Finally, in Chapter 10, I look at the coordination of environmental policies through contemporary funding practices, focusing on gardening the city. The chapter uses a notion of ecologies of action, or the multiple modes of ordering that are necessary to get things done, to argue for a reconfigured understanding of environmental sustainability.

First things? Nature and the sciences

Some images: cows unable to stand, falling onto hard concrete milking yards. Farmers worried for their animals, and for their livelihoods. Government ministers desperate to reassure a doubting public that British beef was safe to consume. Young adult patients, confined to beds, no longer able to voice their suffering, unable to feed themselves or coordinate their limbs. Such images are associated with a major event in the histories and geographies of nature, of food, of agriculture. In Britain, at least, it will be a long time before scientific and government assurances concerning food and medicine safety are taken at face value – from GMOs (genetically modified organisms) in food, to MMR (measles, mumps and rubella) in childhood vaccinations, the links between society and nature, between culture and things, have become manifest. Perhaps for more people than ever before, BSE produced a world where all manner of matters crossed from animals to people, where science dealt in uncertainties and probabilities rather than 'facts', where government, science and business were no longer regarded as separate or separable enterprises.

There are many ways of discussing BSE and many facets to the issue. In this chapter I want to dwell on the relevance of the discussion in Part I to argue that we need to understand things like BSE as multiple rather than single matters. The point can be summarized by saying that BSE is made up of a variety of practices, from protein folding to animal rearing, from recovery of fats in animal rendering processes to a civil servant's memoranda of meetings. As a result of different practices in different places, BSE will not be one single, smooth thing. It has numerous connections, some of which will be more important than others, but most of which will be relevant to how the disease takes shape(s). In other words the disease has geographies, and these geographies will also be in the making, in process. The consequence is that the coherence of BSE will be always a matter of continuing assemblage. How to deal with this multiplicity, with this geography, is a question which motivates this chapter.

The chapter is divided into four sections. I start by discussing the nature of the disease, or more correctly, the multiple, spatial natures of BSE. In the second section of the chapter, I trace how these spatial natures were ignored or downplayed in order to secure, in quick time, an agreeable policy community. Arguing that the recourse to natural-born decisions was a mistake that in turn produced new disease events, I go on in sections 3 and 4 to discuss the kinds of political and other groupings that might be more attuned to this multiple spatiality. At various points in the chapter I point backwards to and extend the argument in Part I of the book in the light of these empirical and political questions.

First things – conforming proteins in practice

As we saw in Chapter 3, towards the end of the twentieth century a heresy was brewing in biology. A group of clinically similar diseases, called transmissible spongiform encephalopathies (TSEs), which included scrapie in sheep, Creutzfeldt-Jakob disease (CJD) in people and Bovine Spongiform Encephalopathy (BSE) or so-called mad cow disease in cattle, was being transmitted or passed from organism to organism by infectious proteins – which eventually became known as prions (itself derived from a slightly odd acronym, proteinaceous infectious particle).

Why was this a heresy? Despite the etymology (protein means primary or first thing), it is most unconventional to think that proteins cause things to happen. In a world divided into those that do (have agency) and those that are done, proteins fall into the latter. They are products not producers. Within the discourse of the central dogma of molecular biology (see Chapter 3), the causes of transmissible disease would need to be able to replicate themselves in a host and thereby carry the molecular information necessary to do so. Endowed with *in*formation, *in*tent and agency, these nucleic acid carrying (id)entities are able to write the disease script into previously non-diseased spaces. Yet, in the case of transmissible spongiform encephalopathies, infectious materials produced in laboratories and reproduced within laboratory animals seemed to contain no nucleic acid. This was a disease that was replicating without the necessary machinery.

How could this material replicate and transmit disease without the conventional source of information to do so? How to be lively without the conventional prerequisite for being alive? The same question has been asked of viruses for some time – how to be alive without cell structures? For viruses it became conventional to explain their liveliness through their possession of sequential information. The arrangement of bases in their nucleic acid structure allowed them to have identity, and to move between hosts, using the hosts' cellular machinery in order to effect replication. But prions made this informational model of an entity or identity problematic.

Prion identity was not a function of sequential information. Once researchers had managed to produce a bio-molecular picture of proteinaceous infective material, a task that involved laboratory workers mapping the sequence of amino acids contained in the protein, prions turned out to be identical, in this informational sense, to other proteins that existed in mammalian brains. The question then arose as to how could seemingly identical molecules be benign and malign at the same time? One answer to this problem was produced by expanding what was meant by biological information. Attention was effectively turned away from molecular biology and towards biochemistry, and thereby to the molecule's conformation, its spatial relations.

Despite the heretical status of replicating proteins, the idea that seemingly mute molecular assemblages could be lively was less surprising in those areas of biology which were used to more relational ways of thinking their objects (prominent and popular examples can be found in the works of Richard Lewontin (1993), Brian Goodwin (1994) and Steven Rose (Rose, 1998; Rose et al., 1984). As Rose (hardly a social constructionist) has it, proteins have more than one identity:

> Each protein consists of a unique sequence of several hundred amino acids. The sequence is known as the *primary structure* of the protein. But the chain is coiled up in helical and pleated patterns, wound back on itself into a configuration which is held into shape by complex arrays of electrochemical forces (these patterns and arrays are known as the *secondary* and *tertiary structures*). Within this globular mass are trapped other, smaller molecules and ions – hydrogen ions derived from water, and metals including calcium, magnesium and iron. Deprive the protein of these smaller ions or molecules, or shift the acidity or alkalinity of the solution in which it is dissolved too far from neutrality, and the globular structure collapses, often irreversibly – this is what happens when milk curdles, for instance. Furthermore, proteins in living cells do not exist in isolation. They are linked with other proteins into higher-order (quaternary) structures, or embedded in lipid membranes, or tightly bound to RNA or DNA. So how do we define a protein? By its primary sequence, or its tertiary structure in space? Do we include all the ions and molecules it collects around its surface and within its crevices? What constitutes the Platonic essence of the protein – or is there no sensible way that we can ask this question? (1998: 39–41, original emphasis)

Rose goes on to ask whether proteins can be defined by functions rather than form. Again though, things unravel, as several types of protein can serve the same function and there are parts of proteins that are more functional than others. The broader point is that definitions for Rose can only be operational and are thereby dependent on the purposes at hand. In other words, definitions work in some situations, only to become redundant or hinder understanding in others. 'A protein is no more a clear-cut natural kind than is an organism or a species' (Rose, 1998: 42). What this amounts to saying is that rather than simply look for discrete objects, that are unchanging and unchangeable, we also need to understand things, even first things like proteins, as matters in relation, as in process. This is not the same as saying that proteins are figments of the imagination, or ideological constructs. Rather, they are things with a geography that includes their relations with all the other elements and molecules that Rose mentions as well as with laboratory scientists, biologists, and much more besides. So things are spatial, they are made in relation and can only exist, can only do things, can only act, can only be defined in relation to other things. In sum, biologists, it seems, like chemists, work with atoms and molecules 'that vary considerably depending on the form and circumstances of their associations with others' (Barry, 2005: 56; see also Bensaude-Vincent and Stengers, 1996).

Following this line of thought, and according to the prion hypothesis, rogue prions (the malign versions of the proteins) are characterized through their relations rather than by an essence, substance or sequential structure. That is they work, or contribute to the production of disease, through their complex of relations which cause them to fold differently to normal prions. These differently folded prion proteins replicate by forming relations with normal prion protein, producing what chemists call a heterodimer (a double molecule, formed from unlike kinds) which then dissociates into two rogue molecules (and so on). The abnormal protein then characteristically binds with other rogue forms to produce plaques that eventually grow to interfere with brain functions.

The main author of this prion hypothesis, Stanley Prusiner, has called prions 'infectious agents' and talked of 'prion replication' (Prusiner and McKinley, 1987). But there are those who would mark a strong boundary between the mechanisms hypothesized for the prion and those that pertain to what are considered to be more bona fide replicators (the viruses and bacteria that we mentioned briefly in Chapter 3). Here is one way of making a distinction between true replication (attributed to viruses, etc.) and what the authors would regard as a lesser form of replication:

> [Prion replication should not] be understood in the same way as virus replication. In the latter, virus infection of a cell leads to the synthesis of exact copies of the virus; in the case of prion replication, the infecting PrP^{sc} [Scrapie, or abnormal, prion protein] causes the host PrP^{c} [normal prion protein] to convert to the abnormal conformation (shape). The newly produced PrP^{sc} molecules need not be exact copies of the infecting PrP^{sc} which may actually have come from another species. (Ridley and Baker, 1998: 137)

There are doubtless differences between these types of replication, but it may be overstating matters to make such a hard and fast distinction. Viruses sit in a hallowed zone for mainstream molecular biology as they are the closest one can get to a naked replicator, to a strand of DNA or RNA. But viruses don't work on their own either. They can sit indefinitely in test tubes as crystalline powders without ever being able to replicate. So even for card-carrying replicators like viruses, the notion that replication is something inherent to an object is strained, to say the least. As Rose puts it '[p]urity in nakedness is sterile' (1998: 253). Replication, for the virus, for RNA, as well as for prions, requires relations and difference. So, returning to the distinction raised above, even though the purity of the copy may be different (although this molecular biology version of pure replication and pure repetition is perhaps rather too wedded to a reversible version of time) the spatiality of process is shared across both forms of infection. Both within the host and circulating through shared food chains, replication is always achieved with others. It is enacted not through any essential characteristic, nor solely as an information-carrying entity, but through spatial relations, through con-formations. It's a case of being with another (conformed) rather than simply being in another (informed).

So what is this brief detour into the contested fields of biology saying? Well, it's starting to say that things are heterogeneous, made with others, they are not defined by an essence – they are, in other words, spatial matters. Following Doreen Massey's (1999; 2005) work in matters spatial, it is also saying that if matters are made in relation, then for anything new to be created there must be difference and there must be space (see Box 6.1).

Box 6.1   On space, multiplicity and openness

For many geographers, space is not something that already exists, it is not the neutral field within which history runs its course. And yet space is often viewed this way, as the unchanging backdrop or stage, upon which stuff happens, or things change. We often say, for example, that things change in time. We rarely remark that things change in space (what we do often say is that as we move across space, say, from one country to another, or from muscle tissue to pancreatic tissue, that things have changed, but this is generally only to indicate that different things are arranged on an already existing spatial plane). But what if things really do change in space? Or to put this another way, what if space is not only the arrangement of difference, but the outcome and possibility for difference? What if, for example, things like countries, prions, muscle tissue, don't just change in time (becoming developed, or going into recession, becoming malign, or ageing, for example), but they also and necessarily have to change in space in order to become these different things? So to become malign, some changes must occur in the spatial relations of a prion protein. To become more industrialized, a country will change its spatial relations with other countries, and within its boundaries. In short, what if becoming and difference are not only about how things alter in time but also and necessarily how they change spatially? And what if these

(Continued)

(Continued)

differences actually make time and space and vice versa? For the geographer Doreen Massey, this kind of argument has all manner of implications, suggesting that we need to think about both time and space, change and stasis, becoming and being, together. More particularly she offers a useful checklist for thinking spatially (see Massey, 1999: 28):

1 Space is a product of interrelations, it is constituted through interactions.
2 Space is the sphere of the possibility of the existence of multiplicity; it is the sphere in which distinct trajectories exist ... Without space, no multiplicity; without multiplicity, no space.
3 Precisely because space is the product of relations-between ... it is always in a process of becoming; it is always being made. It is never finished; never closed.

There is a lot going on here, but for our current purposes it is useful to emphasize that we should refrain from explaining or describing matters as products only of time. To understand matters it is important to consider spaces and times. In particular, it is the coeval trajectories of things that make matters matter and make things happen. Perhaps most crucially for this chapter, because things are made as a result of many other things, they are never fixed. Entities may seem more or less stable, but they are always open. Even if something looks absolutely fixed and solid (like a mountain or rock formation, to take Massey's (2005) example) they are nevertheless in process. They can change.

Once we start to understand biological and other forms of agency as conformational as well as informational, then things start to look less determined. They are made of multiple relations, and thereby are more than one thing, and they are likely to change. Of course, things do tend to hold their shape, and proceed in a fairly predictable fashion. They cohere. But that does not mean that they are closed or singular. In terms of animal health, for example, diseases rarely cross species barriers. It matters for scrapie prion replication that it is conforming with ovine rather than bovine tissues. It matters to a foot and mouth virus that it is with a pig rather than a human. But that is clearly not all, otherwise things would stay as they are. As the case of the human form of BSE made tragically clear, species barriers are not always maintained.

If things are heterogeneous, made with and by others (which are also of course similarly heterogeneous), then the possibility for change is always present, even if it is not always realized in practice (for an indication of this potential in relation to the circulation of animal tissues, see Figure 6.1). There are two ways to understand this potential. The first is to say that things are made of associations and are thereby sociable (Latour, 1999; 2005), the result being that there is a likelihood that they will, in *time*, make other associations and thereby change (see Chapter 3). The second way is to say that things are multiple, that is they are, at any one time, pulled and pushed in many directions and it will be the practicalities of dealing with these spatial and material differences that will start to shape how things turn out and how time moves on (Mol, 2002). In other words, things have histories and geographies.

The result of these two possibilities means that making policy on the understanding that things will stay as they are is to underestimate the potential for change. There are no fixed identities, which means that relations aren't fixed either. Entities with single identities and total networks (which amount to the same thing, closure) underestimate the liveliness of matters. They miss the openness of space-time and the multiplicity of things. As the following sections suggest, this is important for it has consequences for the ways in which policy and politics are understood and enacted.

Making policy without politics: sticking to non-stick facts

Experimental matters are not only being produced in scientific laboratories. Long-term experiments are also taking place elsewhere. For example, since at least the 1920s, animals raised on farms for the purposes of meat and dairy industries have been fed a form of protein-enriched feed that contained material derived from animal carcasses (waste from the meat industry), as part of an intricate circulation and exchange of animal tissues (see Figure 6.1). Figure 6.1 does not claim to be comprehensive, but gives an idea of the different ways in which a carcass could potentially find its way back to live cattle, as well as routes to human use, some less well known than others. The feeds contained rendered animal products, called meat and bone meal (MBM), which were predominantly used in winter months, especially in dairy cows, in order to boost milk production. The meat and bone meal were produced via a process of 'cooking' animal carcasses, adding organic solvents to increase fat recovery and then further heat treatment to remove the solvents (a process that was abandoned in the early 1980s as fat markets declined, and the industrial process shifted from a batch to a continual system). It's a moot point whether or not these changes helped to actualize those prions that could infect cattle and people (prions may well have been able to survive the more exacting conditions prior to these shifts,

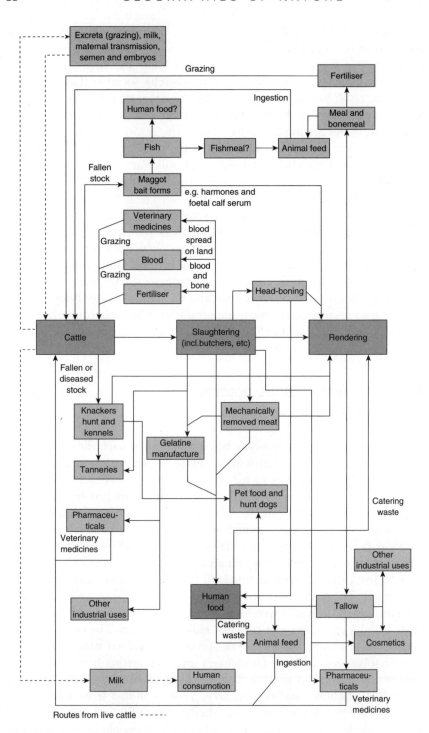

Figure 6.1   The complex routes that parts of the cattle carcass take, often undergoing a series of production processes

and therefore may have been brewing for some time, silently moving through cows and people or, where clinical symptoms flared up, the subject of mistaken identities).

Whatever the conditions that made BSE in cattle a reality, once clinical cases of a new TSE were identified, and the case load started to grow (see Table 6.1), two crucial epidemiological problems arose for the UK government. First, how might the disease be prevented from infecting more of the UK herd and, second, what implications did the disease have for human health? Both problems were addressed in a similar fashion. The approach can be characterized as, first, a search for incontrovertible facts that could furnish a policy, and second, a process of convincing others of the veracity of the facts in order that they enacted the policy. See Box 6.2 on science and policy.

TABLE 6.1   NUMBER OF CONFIRMED CASES OF BSE BY YEAR
OF CLINICAL ONSET, UK, TO 20 MARCH 1996

Year of clinical onset	Number of confirmed cases (UK)
1986	12
1987	460
1988	3155
1989	7807
1990	14736
1991	26041
1992	37545
1993	34231
1994	23225
1995	13934
1996 (to 20th March)	1807

Source: Adapted from Phillips et al., 2000

Box 6.2.   Science, facts and apolitical policy

In what Latour (2004b) calls the bicameral modern constitution, there are two houses, one for Science and one for Politics. The idea is that the former pronounces on the facts of the matter while the latter organizes and implements

(Continued)

(Continued)

the necessary social changes in order to manage the effects of those facts. The former deals, then, with nature, the latter with society. Of course, it follows that if the house of science is asked for the facts, then there is little for the house of politics to actually discuss – there can only be more or less efficient means devised to meet already pre-specified ends. In pronouncing the matter of nature to be fixed, or a closed issue, a matter of fact, then it follows that society and politics must also be closed. The result is that science is effectively a means to bypass politics (Latour, 2004b). Politics is reduced to policy and policy to the technical delivery of natural-born decisions, or the following of fixed orders derived from a determinate matter of fact. The flow of action is like this, science reveals nature (as matter of fact) and parliament then makes policy in line with those firmly established facts. The two houses of science and politics are linked by a single arrow, which allows for a unidirectional flow of apolitical information (see Figure 6.2). The result is that politics is left merely to follow the dictates of science's view from nowhere (see also Chapter 1). As an informational model of action, there is more than a passing resemblance to the central dogma of molecular biology (see Chapter 3).

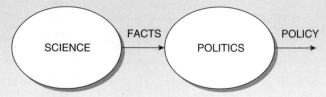

Figure 6.2   The central dogma of policy

Here, briefly, is how this worked once the frightening mad cow disease had gone public.

1 Find the smooth facts, avoid the sticky matters

In May 1988, once BSE could no longer be ignored, the UK government started to establish expert committees and working groups. The most notable of these was the Southwood Working Group. It was set up to report on the implications of BSE in relation to animal health and in terms of any possible human health hazards. In setting about its task, the group enacted a peculiar form of expertise. Rather than drawing on those *experi*enced in the *experi*mental difficulties of TSEs, or those who were *experi*enced in the complexities of agricultural-industrial processes, those who, in other words, were *expert*s in the contexture of their fields, committees were explicitly assembled to include expert scientists who were non-territorial on the matter of TSEs (Lacey, 1994; Pennington, 2000). So, those people and machines involved in the heated controversies over the causes and pathogenic agents of TSEs and those involved in their circulation through food chains, did not sit on expert committees, nor were their experiences given much airing in the work of those committees.

In short, the expert scientists on the committees were dealing from the off with a disease stripped of its multiplicity. In part, this was the result of a government view of expert committees as answering machines, whereas laboratories were acting as question-generating machines (Rheinberger, 1997) (machines that were producing a sticky substance with seemingly no nucleic acid, which was impossible to purify. In other words, the epistemic object in the labs was well short of a non-stick fact – see Chapter 3 and Hinchliffe, 2001; 2005; Keyes, 1999a; 1999b). This mismatch led to relative silence on the uncertainties surrounding the complexities of TSEs in labs. Just as importantly, the strange contextures of infectious proteins in the wild were not discussed in detail. No farmers, abattoir workers, renderers or other industry workers were consulted on their detailed practices. The expert panel's attention was instead focused on the nature of the agent, and following this, on how it could be stopped in its tracks.

So it was the distant, so-called objective, cool headed, uncontroversial or undifferentiated version of science (see Box 6.3) and thereby nature that was considered as expert witness. Rather than link expert to experience and experiment and to a view from whereabouts (see Chapter 1), the two-house system or central dogma of policy-making linked expertise to a view from nowhere, which could coolly adjudicate on the nature of the problem. Nature in this version of affairs is considered as independent of all the huff and puff of politics and conflict, and it takes a non-territorial expert to cut through the noise to access the true facts (see Chapter 1 on this link between views of science and their treatment of geographies of nature). Being objective, in this version of science and policy, requires that the object is smooth. But objects that are pure, that have no links, that are homogeneous rather than heterogeneous, are, as I have argued, dead objects. If, on the other hand, objects are things, heterogeneous and made up of all manner of other things and relations, and practised in multiple and spatially differentiated ways, and can adapt, and accommodate difference, then the meaning of objective might need to change. It might need a different kind of knowing than that performed by the panel.

Box 6.3   Controversies, conflicts and differences in science

In science we are used to there being controversies and conflicts. Scientific debates are often presented as such in the mass media. More generally we have become used to characterizing differences in scientific knowledge in this way, for it tends to presuppose that beyond the controversy there is a single objective truth to be found. This perspectival politics (see Chapter 1) suggests any disagreement is a matter of a temporary controversy which will be solved by matters of fact, or explained away with reference to different theoretical approaches. Another species of this form of analysis can be found in some social studies of science where such controversies are turned into conflicts of (social) interest. Annemarie Mol summarizes the philosophical and social science approaches to scientific difference nicely: 'Where philosophers of science were concerned with the way logical contradictions were handled in practice … sociologists said that 'logical' contradictions do not exist outside the practice in which they are defined. Social conflicts generate contradictions' (2002: 91). What interests Mol, however, is the possibility that contradictions and controversies exist only when closure or resolution is attempted, and are therefore rare events. At other moments, different practices can co-exist. The co-existence of difference becomes even more apparent when our view shifts so that it is not just scientific researchers who are doing things but other practitioners as well. The point is that these various practices may not contradict one another but are, rather, diverse ways of handling reality. In this sense, ontological politics is not only about conflict but also about living with difference, living geographically. So, 'we have left the sociological focus on conflict without shifting back into the philosophical fascination with logical contradictions'. Instead, we are in a place where 'clashes may occur – or different ways of working may get spread out over different sites and situations, different buildings, rooms, times, people, questions' (ibid., 13–14). Drawing on the political theory of Mouffe (1993), this is a place where difference is taken more seriously (see also Massey, 1995). However, unlike Mouffe, this world of difference is not limited to human political subjects. It is also a world where nonhumans, proteins, animals, rocks, can and are multiple and live with internal and external difference.

To be sure, the working party had to deal with massive complexities in simply being able to make some sense of the new disease, and some of their work was crucial in dealing successfully with aspects of the crisis. Through skilful epidemiological work they identified meat and bone meal in animal feed as the possible vector for disease transmission. But famously they tended to underplay the uncertainties regarding the future sociability of the disease. They concluded that BSE was scrapie in cattle, and as a result the chances of the disease crossing to people would be remote (scrapie had, after all, been around for a long time, with no known effects on people who ate sheep meat) (Southwood, 1989). This highly feasible link between the diseases did not turn out to be correct. More importantly though, it was the failure in the report to express the uncertainties surrounding such statements that was the most cause for concern (like, for example, the suggestion that if the disease had jumped from sheep to cattle, then why not from cattle to other species?). A committee that had actively sought differences, that had treated the disease as multiple and produced in practices, may well have found it more difficult to produce such a calming document and more able to counter the repeated attempts by government ministers and officials to downplay the risks and uncertainties surrounding BSE.

2 Stick to the smooth facts

Once a commitment to smooth facts is made, it is difficult to undo matters or acknowledge uncertainties, without losing face. From 1988, when policy started to be formulated, to 1996 when the UK government admitted that there were human cases of BSE, government officials made pronouncements that were consistent with the notion that the facts were 'in' and that matters had been settled. So, for example, when it was announced that certain parts (see Figure 6.3) of even seemingly healthy cattle should not be allowed to enter the food chain, government officials and ministers were asked if this Specified Offals Ban would ensure that British cattle were safe to eat. The response was as follows. Officials and ministers told concerned consumers that the disease was in a dead end host, and would not therefore cross to people. The offals ban was a precautionary measure to ensure that any BSE containing tissues were removed from the food supply to people and to animals. Laboratory trials had shown it was these parts of cattle where the disease resided. These government assurances neglected at least two issues:

1   The infectivity trials in question had been undertaken to identify a disease agent and not to provide public safety information. The former infectivity trials worked with the most concentrated and efficient transmission materials, and said very little concerning the transmissibility of TSEs through less effective tissues (Ridley and Baker, 1998).

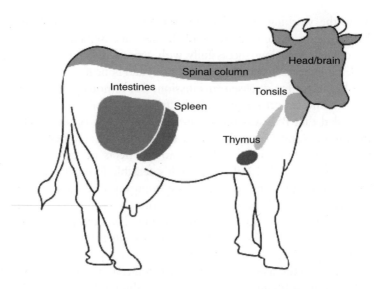

Figure 6.3   Map of a disease – idealized map of specified bovine offal, material that was required to be removed and prevented from entering the food chain from all cattle over 6 months old. Specified offal included organs that were thought to contain the highest concentrations of infective material. One problem was that no one knew if this meant the rest of the animal was safe to eat. Another was that, on the disassembly lines, organs are not always so neatly bounded or dis-connected from the rest of the animal. The latter issue was compounded by the procedure of the mechanical recovering of meat from the spinal cord – a process that could result in specified tissues being incorporated into the food chain.

2   Assurances played down and/or failed to consider the complex processes involved in killing and disassembling cattle and the effect this would have on the ability to declare that specified offals would not enter the food chain. It was indeed a questionable assumption to suggest that these specified materials could be removed cleanly on an animal disassembly line. The speed of the process, the stressful methods used in the killing of animals, the use of mechanical saws in the butchering of those animals and the practice of mechanically recovering meat from the spinal cord would all blur the boundaries of safe and unsafe tissue.

In a similar 'decide, announce, defend' mode of operation, on being asked why a ban on using meat and bone meal in feed for ruminants had not been extended to its use in fertilizer (which might be eaten by cows grazing in fields) or to feed destined for export and non-ruminants, government officials told renderers and feed manufacturers that a massive dose of infective material was needed for the disease to strike – so MBM containing fertilizer dispersed in pasture would be safe and feed abroad was largely used for non-ruminants, and any small amounts ingested by ruminants would have no effect. The story was effective; it secured agreement and a policy community in broad agreement that there was a need for the ban – but at a cost (Hinchliffe, 2001). The policy frame, or the terms of reference that officials designated as important to making policy, did not include the possibility that prions could be highly effective even in minute quantities and that the architecture of the ban would still lead to small levels of MBM in ruminant feed. The development in the 1990s of BSE in cattle born after the feed ban provided a stark reminder that the BSE collective was far from enclosed by the policy frame (see Box 6.4 on collectives and frames). Cross-contamination in feed mills had seemingly prolonged the disease, and possibly assisted its spread to humans.

Box 6.4   Collectives and frames

A *collective* is an unfinished accumulation of people, machines, animals, ions, amino acids – in short an (ongoing) assemblage of nonhumans and humans (see Chapter 3 on assemblage).

A *frame* is a more or less temporary means to define what is in and what is out of an account or of a shared term of reference. The notion comes from Goffman (1971). In many cases the act of framing comes after a collective has started to form, and some kind of ordering has been enacted. Frames tend to bracket rather than abolish the world outside, and in ranking matters in this way they can have effects (Callon, 1998a: 249). For example, they can place some matters out of view, thereby ignoring their proliferation, or they can focus too much attention on those matters inside at the expense of those outside. Frames will therefore tend to have an effect on the continuing make-up of a collective.

There are several points to make here which will be expanded upon in the next section:

- Participation of so-called non-expert or lay groups (like the questioning public, farmers and industrial meat and feed processors who were asking awkward questions) in policy making means very little if single versions of materiality (smooth facts) are the only species of thing up for discussion. Everything is reduced to a controversy that is closed through a logical argument. In this case, questions from industry and consumer groups were persistently deflected by matters of fact rather than taken seriously as indicators of multiplicity and difference.

- Facts of the matter shield practices and thereby other materialities from view. The notion of smooth, incontrovertible facts underplays the ways in which matters are made in and through practices. The 'massive dose' fact excluded a discussion of the problem of cross-contamination of feeds in feed mills and on farms.
- Frames tend to spring leaks. There will tend to be overflows (Callon, 1998a). Specified offals did not fall neatly off the disassembly line, just as ruminant-derived meat and bone meal were not confined to the bags labelled feed for non-ruminants.
- These overflows are further enhanced by ignoring their possibility. To use Latour's language, the allocations of matters of concern to matters of fact, of quasi-objects to matters of nature, secure a secret breeding of new associations (Latour, 1993; 2004b).

So how to deal with conformations and overflows?

Ways of being open

Here are two versions of how to make policy in the light of animal and human health concerns. They were aired at the UK government-backed inquiry into what had gone wrong in the case of BSE. The inquiry was extensive, sitting for over 18 months and producing a 16-volume report in 2000 (Phillips et al., 2000).

Kevin Taylor, Veterinary Head of Notifiable Diseases Section, MAFF, told the BSE Inquiry that:

> In animal control terms it is perfectly normal to put in place what seems to be a logical and simple control. There is no virtue in going for complexity, it is better to go for simplicity, and to observe what happens. When you get evidence that what you have done is either unsuccessful or could be improved, if you find evidence that it could be improved or that it is failing in some way, then you look again at the situation and you make the changes which are necessary to block the leakage, if you like. It is like blocking a dam with stones. It is easy to block a big part of the transmission and it is only later that you become aware that it has been leaking a bit round the edge, so then you block the bits round the edge. That is exactly what was done with BSE and it is exactly what historically has been done with other animal diseases. (Phillips et al., 2000, vol. III: 111)

This sounds practical. Agree the main problem, and then fingers crossed that nothing has been missed (or that any overflows won't amount to much). But one obvious drawback to this approach is that problems may refuse to present themselves – and it often takes a lot of work to make an overflow *present*. Prions don't always announce themselves or come ready made into view. Indeed, the long incubation period for this disease, taking several years to move from infection to clinical presentation, means that leaks become apparent a long time after the event. To make matters worse, there was no field test which could detect meat and bone meal in feed, and there was no way of testing for sub-clinical BSE in cattle. So things may well have become catastrophic before anyone had noticed that there were leaks. Given these kinds of problem, here is a response and recommendation from the BSE Inquiry team:

> Failure to address the risk of cross-contamination of ruminant feed with non-ruminant rations resulted from a failure to give rigorous thought to the question of dose. The best time for rigorous consideration of the practical implications of a set of Regulations is at the time of their introduction. Once Regulations are in place, there is a tendency not to consider potential problems, but to wait to see whether any arise in practice. The question whether cross-contamination would be a problem was an obvious one to anyone with knowledge of how feedmills operated. The question was raised, but dismissed by Mr Meldrum [Chief Veterinary Officer 1988–1997] without the rigorous consideration that he should have seen it received. No one could know the minimum quantity of infective matter necessary for oral transmission. In the absence of knowledge, the ruminant feed ban should have been implemented on a 'worst case' assumption. (Phillips et al., 2000, vol. III: 117)

The BSE Inquiry team are right in that once a policy is made, there is a tendency not to consider problems until they arise. In other words the matter is treated as closed until we hear otherwise. So, as an alternative, it is better, they argue, to do everything possible to head off problems at the outset in order to minimize the risk of something cropping up later on. This kind of thinking has been characteristic of a shift in administrative style in the UK, post-BSE. The shift was, at least in intent, from expert closure to more open, transparent policy-making, which can be more attendant to the different knowledges (sometimes) and materialities (rarely) that are implicated in any attempt to form a frame of reference (Hinchliffe and Blowers, 2003). It is worth noting that the shift was very subtle, geographical, and subject to review. Decisions taken in Westminster are rarely reopened completely. From staging 'open' discussions of genetically modified food (little more than public relations exercises) to the disclosure in 2006 that any enquiries into new nuclear power plants in the UK will be partially foreclosed by pre-licensing agreements, ruling out any discussion of the design of nuclear plants, openness is circumscribed and tends not to extend to materialities, and does not seem to travel very far from government offices in the capital.

And yet, apart from offering a seemingly more inclusive framing of a policy, the Inquiry team are not so different from the dam builders (represented here by the previous quotation from Kevin Taylor). Both versions of policy making assume that a frame is the norm and that overflows can be managed. As Michel Callon has argued, such an attitude, which is common across administrative and economic sciences, prioritizes closure in order to avoid premature overflows (Callon, 1998a). It leads to policies of containment. An alternative, which is more akin to the sociologies of association and certain kinds of geography and biology that have informed the geographies of nature that were introduced in Part I, and which I traced in the discussion of prions, starts from the assumption that overflows are the norm and any framing is a hard-won achievement, one that requires continual work in order to maintain itself. Which means that just as the strange folding of chains of amino acid can be thought of as produced through the molecular relations which make the prion protein possible, so too is a policy, like the ban on meat and bone meal for ruminants, conditional upon and subject to other possible relations. Callon puts it like this: Every element involved in a frame, 'at the very same time as it is helping to structure and frame the interaction of which it more or less forms the substance, is simultaneously a potential conduit for overflow' (ibid.: 254). So hydrogen ions in normal prion protein are both of the protein and part of the actualization of rogue prions, and the feed ban contains elements that form many other relations besides. Feed manufacturers were not only producing feed for ruminants in the UK. They were manufacturing lots of other products, including feed destined for pigs and poultry, and feed for export overseas. Machines that make feed pellets for cattle were also therefore involved in other markets and with other animals. Some of this might have been anticipated (and internalized into the policy frame), but the point to stress is that things relate not just to a current frame of reference (however narrowly or widely it is drawn), but to lots of other matters too. They are, in other words, more broadly related than the frame would suggest. And the point can be extended by saying that not only do relations cross frames, but that things are also likely to be members of more than one network in ways that may produce conflict or may simply create the possibility for new configurations to develop.

The result of Callon's line of thinking is that any policy, conceived as a frame, will spring leaks. Any attempt at ordering, at forming boundaries, is subject to overflows (Callon, 1998a). Another way of saying this is that the production of order also produces, at the same moment, disorder. And this facility of order to create its other, disorder, is enhanced when the ordering strategy includes a claim to absolute truth, to being in the nature of things, to being incontestable. In other words, suggesting that the ban was of nature rather than a complex crossing of nature-culture produced the material conditions for further disease events. Purification of stories helped to brew a hybrid, a crossing, a quasi-object (see Box 6.5).

Box 6.5   Divisions and proliferations – the gestation of quasi-objects

The more that nature and culture are forced apart, the more that a space is generated for the illicit making of things which don't fit. The more we try to purify things into human and nonhuman, the more we manage to accelerate the production of things that refuse to fit, that come back to trouble our purification schemes. This is one of the main claims of Bruno Latour in *We Have Never Been Modern* (1993). To strive to be modern is, for Latour, to believe that we can hold science and politics apart, that we can proclaim the truth of the matter, that subjects and objects are separate entities – something that pre-moderns found impossible, given their propensity to mix animals and politics, law and demons, spirits and truth.

Yet, as we forbid ourselves the possibility of understanding science and politics together, we inadvertently nurture further crossings (Bennett, 2001), or quasi objects as Latour calls them (Latour, 1993), matters that are neither reducible to culture nor to nature. And, given that as moderns we tend to celebrate purification while denying these crossings (or at least treating them as open secrets (Bennett, 2001: 96)), these hybrids are often shielded from view, so much so that they proliferate, haphazardly. Until, that is, they erupt onto the scene.

So we might say that (contra the quote above from the BSE Inquiry team) the best time for rigour is not only at the introduction of a regulation, it is before, during and after, until such time as matters of nature have settled. And that what we mean by rigour and objectivity might need to change. Rather than an all-seeing policy maker, we might instead redefine rigour and objectivity as more akin to an experimentalism that is prepared to accept that things can change. The point here is not to consider nature as the starting point for policy or politics, but as a fragile end point, something to work towards rather than spuriously refer back to. It is not the origin, or foundation, or bedrock which somehow grounds or secures the right actions. It is, like those actions, the fragile outcome of a vast array of other activities and things.

The politics of things

So what is this saying about a politics of things and geographies of nature? There are at least three things to state. First, because things are multiple, because they overflow, because they are made up of things that are conduits to elsewheres and to other things, then there is always going to be a struggle to keep things together. Unlike the smooth facts that furnish some versions of the world, matters are sticky, and will associate more or less widely. Similarly, unlike the Third Way politics of Anthony Giddens (1998), and seemingly most western social democratic political parties, natural matters do not sit in a neutral space which can transcend political divisions. 'Environment', 'Nature' and other such terms are not the uncontroversial good things that can unite a fractured social world. As I have shown, matters of fact are not the pre-existing centre ground that can bypass politics (Latour, 2004b; Latour and Weibel, 2005). Nor are they uncontroversial outcomes of deliberative procedures, conflict resolutions, and so on (Mouffe, 2000). There is then neither a pre-existing nor post-rationalization place that can foreclose conflict. Where conformations take shape, they do so through internal and external difference that may well unfold, or change matters. There is then no stable ground, no first thing. Politics includes things, but cannot take those things for granted or treat them as the means to avoid the hard work of gathering together.

Second, a politics of things will be about more than exclusion of others. If things are the outcome of multiple practices, then it is not necessarily the case that these will be in conflict and need resolving. It may be that they are divergent ways of handling reality and need to co-exist in some way or another. The result is that things, whether they are proteins or policies, can endure because they retain the possibility for responding to new conditions not with a lurch toward exclusion, but to an accommodation or subtle variation. In order to affect this multiplicity, things may benefit from being loosely connected (Law, 2006a). It may also be the case that in order to be capable of adaptation, of coping with the new – and we can think of doctors moving from Amsterdam to rural Zimbabwe, (see Chapter 5 and Mol and Law, 1994), or feed ban policies, or H5N1 viruses – this requires multiplicity. In practical terms, this would suggest that a looser gathering of expertise in order to reduce the likelihood of the spread of disease would probably do better than a policy measure that allows for only one kind of expert (and thereby singular materiality). Moreover, it would need to be a gathering that adapted to new developments, that was capable of responding to new conformations.

Third, there are consequences to this realization that things, from proteins to feed bans, are born of multiple relations, some of which involve struggles, partial accommodations, withdrawals, and so on. One consequence of this openness is that there is no possibility of exhausting things in terms of reducing them to substance or to their total relational situation. Things cannot be fully accounted for, described, told or calculated. Which means even relatively settled matters can be capable of the new. Even dead and buried proteins can overturn scientific logic, and can turn from benign to malign matters. Another consequence is that there are lessons in this for the task of re-generating collectives, or re-assembling the social (Latour, 2005). In making frames, attention needs not only to be given to the processes of partial exclusions which might come back to haunt those inside (see, for example, the political process envisaged by Latour, 2004b), attention might also be paid to the quality of relations that exist within and between things. As indicated above, producing a disease response that highlights and draws on rather than ignores the combination of skills, knowledges and techniques that various members, from people to hacksaws, have at their disposal would generate more adaptable but nevertheless robust policies. To be sure there is nothing easy about these gatherings, they still need to make agreements on what is to be done, they need to act and they need to agree on whether or not it is justifiable to take this action rather than that one, but the point is that a disservice is done if the emerging collective's in-built sensitivities are obliterated by issuing a bald order that requires no further discussion, watchfulness or understanding. Constituting such a collective without the short-cut of natural necessity is extremely difficult, but it is a necessary prerequisite to producing a collective that is both solid enough to act and loose enough to be vigilant.

In knowledge terms, such collectives benefit from a looser kind of sense, a knowledgeability that does not claim to know once and for all, and that admits to and engenders multiple practices and more than one ontology. It is to say that the officials' view is not the only view, and that the disease is enacted in many places, from ripped bags of feed in farm yards to high water pressure hoses that are used to mechanically recover meat. And that all these knowledgeable practices are relevant and can be used to gather together a more effective policy community. This looser kind of sense requires a looser form of gathering, which does not claim to have solved the problem of collecting. It requires, perhaps, not so much a natural contract (compare with Serres, 1995b) but an acknowledgement and attention to a society of experience and experimentation (producing a different sense of expert than was enacted by the British Government and the Southwood Working Party). Experience of both the openness of things, their associations and their multiplcity, requires an acknowledgement of uncertainty, a knowing of indeterminacy (Mazis, 1999; Hinchliffe, 2001). This knowing was lost when the central dogma of policy declared the facts of the matter and thereby ruled all other knowledges out of court.

Conclusion

How to imagine these looser ways of sensing, these less determined collectives, or assemblages as Deleuze called them (see Chapter 3 and Law, 2004a)? One way of doing so would be to say that they might have more in common with the dry stone walls of the British countryside than with some firmly cemented dam. The elements or irregular shaped stones will not lend themselves fully to the structure, there are gaps and pieces that jut out, but the result is hardly ephemeral. It is made through partial connections and withdrawals – and adapts well to variations in climate. Meanwhile, constructing a dry stone wall is a highly skilled, experimental procedure, placing and replacing stones so that the structure takes shape. It takes time to learn how to do this, and part of that learning is about getting to know how stones can be placed together in sympathy to their irregular configurations. As a conformational action, it's learning to work *with* the stones rather than *on* the stones.

The metaphor has been used elsewhere to describe a society of experiment. Drawing on Deleuze's work, John Rajchman summarizes thus:

> Deleuze pushes the experience or experimentalism of thought into a zone prior to the establishment of a stable, intersubjective 'we' and makes it a matter not of recognizing ourselves or the things in our world, but rather of encounter with what we can't yet 'determine' – to what we can't yet describe or agree upon, since we don't yet have the words … which allows us to go beyond our social identities and see society as experiment rather than contract. The problem of 'experience' or 'experiment' … becomes one of forging conceptual relations not already given in constructions whose elements fit together not like pieces of a puzzle but rather like disparate stones brought together temporally in an as yet uncemented wall. (2000: 20)

This chapter might have helped to convince you that taking orders from the natural world is not as straightforward as it sounds (see also Part I). One way of saying this is that there is no full access to a natural world which furnishes us with all the information about its potential linkages, about what it might do in all circumstances. The lessons for making policy are difficult ones – no one can pretend that uncertainty is an easy matter to comprehend when decisions need to be taken and politicians need to be made accountable for those decisions. But false assurances are very often more damaging, in their attempted foreclosure of concern, than a purposefully maintained openness. To be clear, this is not a politics of anything goes. I hope that this chapter has also started to suggest that despite being unable to find unequivocal orders from nature, natures still matter. Subsequent chapters take this problem forward.

Background reading

The requirement to keep science matters in the open, so to speak, has been wonderfully conveyed by a whole series of works that have had impacts both academically and practically. Alan Irwin's (see, for example, Irwin, 1995) work is a very good introduction and deals with the early events in the BSE crisis as well as a number of other now classic studies in science and society. A good source for thinking through issues of framing, with some links to the BSE issue, can be found in Michel Callon's essay on framing and overflows (Callon, 1998a). A basic introduction to the BSE issue can be found in Hinchliffe (2000a).

Further reading

For more details on some of the issues raised here in terms of BSE, see Hinchliffe (2001). A very useful introduction to some of the scientific debates can be found in Ridley and Baker's book (1998). The report on the BSE Inquiry is a wonderful resource for understanding the disease and UK government practices (Phillips et al., 2000). For more discussion on the problem of re-assembling without recourse to the Nature/Society dualism, see Latour (2005). For a wonderful example of the importance of working through what can be involved to develop working natural relations in a postcolonial register, without a single Nature, see Helen Verran's work on practising fire ecologies (Verran, 2002).

Securing natures 7

Introduction

In the previous chapter I suggested that the spaces of disease are as conformational as they are informational. The 'thing' about disease is not only the infective particle, it is the relations between various matters that make a disease. I now want to explore this conformational approach with reference to security and nature, or more specifically to the issue of bio-security. The question that animates the chapter is, how can one secure the nonhuman world? The answer, as you may well have guessed from previous chapters, is not to attempt to define pure spaces, or erect impermeable barriers between the pure and the impure. Rather, it is to understand and work with the multiplicities of things, and allow for some 'give' in spaces for nature. The chapter builds on similar theoretical material to previous chapters, but it also extends the discussion by exploring the ways in which things, including viruses and other matters, maintain their shape not through excluding others but through a partial openness to other matters. It is this ability to maintain themselves while adapting to changes in their world that can be important and has implications for understanding and enacting spaces for nature. The implications are twofold. First, separations of Nature and Culture are doomed to produce more rather than less crossings (a point that has been introduced in earlier chapters and relates to Latour's critique of the modern constitution). Second, I argue that any attempt to be modern, any attempt to order nature's spaces, will itself already be impure, heterogeneous and practised in many places, with many things and by many different kinds of actor. The result points to a need to understand the ways in which various activities work for and against each other, interfere and overlap. Echoing the conclusion of the previous chapter, it may well be that rather than building high walls and large dams to control natures, disease 'control' may be best served through a loosely coherent programme which makes space for difference.

Biosecurity

Biosecurity is a term that has started to make its way onto various agendas. H5N1, or the highly pathogenic avian influenza virus, SARS, or severe acute respiratory syndrome, anthrax attacks in the wake of 9/11, foot and mouth disease in the UK, introductions of invasive plant and animal species – all figure a world where various bio-matters have become matters of concern. The associated attempts to secure the bio, to make matters safe, speak to a world where matters biological are also matters political: from beagles wandering over the international arrivals baggage carousels at Australian airports, sniffing for traces of illegal movements of biological material, to mass killings of animals in Thailand, Vietnam, China, Egypt and the UK, where different kinds of animal disease were being 'contained'; from real-time PCR (Polymerase Chain Reaction) devices being tested in US postal systems as possible means to warn of a recurrence of the postal network being used to distribute pathogens, to stock piles of antiviral drugs being procured by the richer nation-state governments – the attempts to secure the bio are also suggestive of something of a panicked world, working in vain to attempt to keep natures at bay and ensuing social orderings intact. These and other movements, mixings and securitizations of bio-matters suggest a world where two things are, paradoxically, happening at the same time. First, a world of sovereign states, of uneven wealth, of North versus South, of militarized and policed boundaries, is being reinforced just at the time when, second, there is every indication that bio-matters and their movements are making such distinctions and demarcations less and less salient. The result is something of a heady mix for those interested in nature politics, and in geographies of nature.

In order to start to investigate further, it is useful to ask, what kinds of bio and what sorts of security are being invoked by the term biosecurity? As we will see, there are many things at stake, so much so that it makes more sense to talk of biosecurities.

Biosecurities

A brief review of the uses of the term biosecurity suggests at least three areas where the bio, or life, is felt to be insecure. First, there are the attempts to manage the movement of agricultural pests and diseases. The focus is often on farm-based practices including emergency measures to contain disease outbreaks and techniques to prevent infection of livestock. While many of the measures are not new, the term and the attendant shift in responsibility for maintaining the disease-free status of a nation's herds onto farmers have been met with some suspicion, especially in the UK where this form of biosecurity was set in train following the devastating foot and mouth disease events of 2001 (Donaldson and Wood, 2004).

Second, there are the attempts to reduce the effects of invasive species on so-called indigenous flora and fauna. For some, human-induced animal and plant movements pose one of the most significant threats to global biodiversity (Bright, 1999). This form of biosecurity has gained a particular foothold in Australasia and on islands where the effects of ecological colonization (the importation, often from Europe, of farming systems and species) have been most pronounced and where the notion of a primordial state of nature is perhaps most marked (Crosby, 1986; Bright, 1999; Clark, 2002). While it is possible of course to detect more than a little 'fortress nature' in this form of biosecurity, with invasive species labelled as culture and compared unfavourably to the so-called indigenous nature of a region, there are never-theless strong arguments regarding the deleterious ecological effects that can be produced by poorly embedded species, like rapacious predators that have few if any predators themselves.

Third, there are the dangers of the intended and inadvertent spreading of biological agents into the human population. Attention here focuses on labo-ratories which handle potentially hazardous organisms, possible uses of pathogens in bioweapons and bioterrorism, and the potential for crossings of animal-borne diseases into the human species (zoonotic diseases).

If biosecurity involves various sorts of things, it also involves many differ-ent kinds of practice. One way of providing a typology of these is to follow Collier and Lakoff's (2006b) identification of three political logics of security. The first of these, Nation-State Security, has its roots in the establishment of nation-states in seventeenth-century Europe, and involves the will to secure territorial sovereignty. It is premised on a bipolar world of friend and enemy, the spatialization of that world into territorial units and a militarization of various borders. The latter can be external or internal (surrounding the enemy within), with the result that policing borders becomes a matter of foreign and domestic policy. Second, there is the logic of Population Security. With its origins in late nineteenth-century social welfare reforms, population security involves attempts to improve overall and individual levels of and access to health and welfare by organizing the purchase of health and social securities. The collectivization of health and social safety nets through public and private insurance and through the organization of health and social infra-structures is the main recognizable mode of this form of security. Finally, for Collier and Lakoff, there is Vital Systems Security. This is a mid-twentieth- cen-tury response to the development of extreme emergencies, including most notably the possibility of nuclear attack or other events which are regarded as incalculable and therefore largely uninsurable (earthquakes, hurricanes and so on). It is worth noting that non-calculabilities are not always innocent matters – they can themselves be produced through particular set-ups such as a reduced funding for emergency bodies or a proliferation of accounts which reduce the likelihood of accountability. (See Box 3.1, Chapter 3 and for more detail, Callon and Law, 2005.) Nevertheless, the rationality here is prepared-ness to respond rapidly and effectively through emergency planning and sce-nario testing (see Table 7.1).

These categorizations are useful to set the scene, and it is possible to start to map together the matters of concern (agricultural pests, bioweapons, poorly embedded species) with characteristic security responses (see Table 7.2). And yet, as a reading of Table 7.2 will confirm, in practice, biosecurities are

TABLE 7.1 SUMMARY OF POLITICAL LOGICS OF COLLECTIVE SECURITY

	Nation-state Security	Population Security	Vital Systems Security
Movement of articulation	Seventeenth-century territorial monarchies	Late nineteenth-century social welfare	Mid-twentieth-century civil defence
Aim	Secure territorial sovereignty	Collectivize risks to the population	Preparedness for emergencies
Object	Enemies of the state (internal and external)	Pathologies of the social (poverty, urban unrest)	Potential catastrophes; vulnerabilities of critical infrastructure
Form of rationality	Strategy	Insurance	Preparedness
Examples of apparatuses	Military, border control, intelligence	Public health, education, urban hygiene, social security	Disease surveillance; environmental detection; data mining

Source: Couier and Lakoff (2006b: 5)

not easily separated into single forms or single logics. There are overlaps and interferences that require us to look more carefully at biosecurity operations *in practice*. Indeed, and as I have emphasized throughout this book, things don't tend to be done as a result of logics alone. As suggested in Chapter 1, Pasteur didn't achieve some of the first farm level interventions in biosecurity in the nineteenth century by convincing others of the logic of his ideas. Rather it was on-farm practices that led to the development and translation of anthrax bacillus vaccinations (Latour, 1988). So it is with contemporary biosecurities, a whole range of actors, micro-organisms, states, animals, people, aeroplanes, and so on interfere with one another and make single logics or orders unlikely to work in practice.

There are a number of ways of underlining the point that biosecurities are rarely clear cut. First, in practice, a particular thing may not fall neatly into the categorization of bio outlined above. In the avian flu case, for example, aspects of ecological integrity (imported 'exotic' birds) mix with aspects of agricultural disease (the spread of the disease in poultry raised for their eggs and meat), and with aspects of human health (the risk of infection to those humans living and working with poultry and the wider fears of the development of a human-to-human transmissibble and pandemic form of influenza from H5N1). Second,

TABLE 7.2   BIO-THREATS AND CHARACTERISTIC RESPONSES

Target group/ matters of concern	Conventional bioagencies	Technology/ apparatus
Agricultural integrity, and	International and national	Border control and policing,
purity	movements of livestock, livestock products and foods with potential to carry pathogens (viruses, prions, bacteria). Migrating birds and insects	surveillance, hygiene regulations, vaccinations. Contiguous and mass extermination of animals. Insurance of herds.
Ecological integrity	Poorly embedded organisms ('non-natives')	Border control and policing, surveillance. Extermination strategies.
Human health and welfare	Zoonotic diseases, animals and people carrying pathogens that have potential to produce pandemics. Inadvertent and purposeful releases of bio-hazards	Border control, environmental monitoring, social security, education, detection algorithms, isolation, surveillance, emergency planning.

different security practices do not exist in historical or geographical isolation. In practice, they co-exist with one another, perhaps competing, or in harmony, or even indifferent to one another. So, for example, population security is enacted against the threat of poverty, but there are instances where the fear of public unrest in the face of unemployment draws in other securities, including nation–state security (with the militarization of policing employee strike action, for example) and vital systems security (where contingencies for maintaining services are in place). Finally, security is probably not best understood as a matter of responding to an already defined threat. security (as Foucault demonstrated only too well), is about the public production of norms, their surveillance, regulation and enaction (Foucault, 1977). Thus, to say it too quickly, the three forms of collective security detailed earlier (state, population, and vital systems) are never simply a reaction to self-evident hazards, but are always in part about the constitution of both a particular kind of body politic (self-contained, healthy, alert respectively) and a particular kind of bio-*in*security (fear of the outside, fear of sickness, terror of the inevitable respectively).

Rather than separate threats and distinct logics of security, it is more useful to consider practices, and modes of securing. In using the term modes of securing, I want to evoke something similar to the notion of modes of ordering employed by Law to convey the multiple ways in which the social is enacted (Law, 1994, see also Chapter 10). In particular, it is to emphasize that:

- There is likely to be *more than one* mode of securing in operation in any situation.
- Modes will *relate to one another* in ways that can be mutually supportive, destructive, indifferent, in conflict, co-exist, reduce each other's effectiveness, and so on.
- Each mode is practised and thereby marks an imperfect *attempt* at security. Therefore, rather than call them modes of security, they are modes of securing. The resulting state of security will always be, more or less, provisional and subject to change.
- They are heterogeneous and distributed over a wide array of actors and things, and don't have a thinking human subject at their centre. They are not therefore equivalent to logics or rationalities.
- The potent mixings and interplays within and between people, places, animals, forms, chemicals, embargos, cells, and so on are more than likely to be generative, to produce new conformations.

Focusing on modes of securing and their practical enactments points to a situation where the question is not so much what kind of security is being rolled out, but how do the various orderings interact with one another and with many other things besides? How do they interfere? In order to explore bio-securing I engage with two examples: foot and mouth disease and avian influenza.

Foot and mouth disease

In a rich series of articles John Law explores the complex practices of foot and mouth disease which flared up in the UK in 2001, offering accounts of the numerous ways in which foot and mouth disease was enacted (Law, 2004b; 2006a; 2006b; 2006c; Law and Mol, 2006). I won't do justice to the complexity of Law's concerns in this chapter, but what I want to take from Law is the notion that biosecurity is practised in many different ways. Let me start with an account of the disease.

One way to describe foot and mouth disease would be to call it a viral disease that affects cloven-hoofed animals. The virus stays viable in moist conditions, is usually breathed in by the hosts and then incorporated into cells which assist in the production of more virus. Affected cells then either burst or expel virus which spreads through the host and is eventually excreted in the form of blisters (characteristically around feet and mouths) or through normal emissions of air, urine, faeces, milk, and so on (Donaldson and Wood, 2004).

Describing a disease in this way marks a fairly neat boundary between healthy and diseased animals. Healthy is defined by the absence of foot and mouth. However, things are rarely so simple. For the highly infectious foot and mouth virus is endemic, or ever present, in many animal herds around the world, herds that are also considered in some quarters to be healthy. Indeed, the virus is rarely lethal to its host (although can be so in very old or very young animals). Animals suffer, and can suffer badly (Law, 2006b; Woods, 2004a), but once the disease abates, recovery rates are high. However, in the terms of agriculturists interested in animal productivity, recovery isn't full. An effect of a foot and mouth infection is to reduce meat and milk production.

Even from this brief sketch it is possible to say that the disease is more than a viral infection – it is also a violation of norms regarding agricultural productivity. As historians have documented (Sheinin, 1994; Woods, 2004a), in nineteenth-century Britain and in twentieth-century continental Europe, foot and mouth was an acceptable if annoying condition for many farmers. There were flare-ups and attempts to isolate affected herds but no real attempt to eradicate the disease. In other words, there were differences between herds and between countries, but these were differences in degree rather than in kind. However, once the virus started to be removed from British herds upon the instigation of slaughter policies in the nineteenth century, the differences became more controversial (Woods, 2004b). They started to turn into differences in kind. An international map could now be produced, separating the clean from the defiled, the pure from the polluted, the productive from the subsistent, the modern from the primitive (Law and Mol, 2006). To be developed and advanced was to be disease free, to be productive, to be efficient. It is worth noting in this telling that there is no room for grey areas. Either you are developed or underdeveloped (see Massey, 2005, on this single space of developmentalism). There was also no room for the vaccinated animal (see Sheinin, 1994, on the differences between mainland European and British policy on vaccination in the early twentieth century). The vaccine takes the form of a mild dose of the disease, to which the animal then develops a level of immunity. The trouble is that this does not stop the animal then becoming a carrier of foot and mouth, a state called 'masking' that can exist for up to three years (Woods, 2004b). It is thought that carriers rarely spread disease, but the risks are not known. So countries that routinely vaccinate are viewed with suspicion by the pure. To retain WTO (World Trade Organization) Foot and Mouth Disease-free status it is forbidden to vaccinate animals for foot and mouth. Slaughter becomes the only policy that 'guarantees' cleanliness, and produces differences in kind that are backed up with isolations and regulations designed to halt the spread of disease (see Figure 7.1).

Figure 7.1 UK government biosecurity publicity. After the foot and mouth outbreak in 2001, biosecurity in the UK was firmly located in farm practices, including routine cleansing activities, especially when people, stock or machinery moved on and off farm premises. It also involved form-level precautions regarding the sourcing of animal and bird stocks.

We have quickly moved from a description of a virus, to a map of disease that starts to take in the relations between viruses and cells, animals and agriculture and between nation-states. But orderings are not always so neat. As the previous chapter suggested, there is always potential for overflows, for elements of a network to act as a conduit to elsewhere. For example, in the early twentieth century, Argentinean-produced meat (dead stock, not livestock) could move to Britain despite the foot and mouth status of the Argentinean herds. The continuation of trade was a complex matter, involving British support for Argentinean accusations of US market protectionism (the US imposed a trade ban following disputed FMD outbreaks), British and French virological research on the passage of viruses in refrigerated meat, and British holdings in herds and packing industries (see Sheinin, 1994), but the effect was that a way needed to be found to protect the UK herd from any contamination from this imported meat that may contain the virus. Although the Argentinean meat was for human consumption (and humans were not affected by the virus), there was a potential conduit to British farm animals through the practice of feeding waste food or swill to pigs. As Law and Mol (2006) tell it, attempts to stop this flow involved a new boundary, and a new set of regulations. By the late 1920s, people's waste food needed to be cleaned prior to feeding it to pigs. Cleansing in this case involved boiling the swill for such a period that any foot and mouth virus would be denatured (Law and Mol, 2006; Woods, 2004c). So biosecurity involved the implementation of practical measures on farms in order to prevent circulation of disease. International borders and regulations were not so much drawn as conventional cartographies, following the political boundaries of nation-states, but were enacted on farms. Another way of making this point is to say that cartographic regions and volumes were on their own insufficient to the doing of biosecurity. In addition, various networks needed to be mobilized in order to prop up the regional statement of a 'UK foot and mouth disease-free herd'.

In the UK, foot and mouth had shifted from being a mild and unpreventable ailment to be lived with (Woods, 2004c), to an eradicable disease. A boundary between diseased and non-diseased was therefore maintained through a variety of technologies including surveillance, slaughter of infected animals and the cleansing of animal feed. Producing and maintaining this difference in kind required a lot of work in a lot of different places (on farms, in government offices, in virological laboratories, and so on). Securing involved a rich variety of processes, materials, people, places, animal lives and deaths.

As I have emphasized in this chapter and in the previous chapter, things are not ordered but lend themselves imperfectly to orderings. And, more than this, any ordering also makes a disordering. So a pure herd also makes the impure herd, and biosecurity as a state or condition makes bio-insecurity a possibility too. The foot and mouth epidemic that beset the UK in 2001 led to unquantifiable human and animal suffering and loss. According to many reports, it all started with the failure of a single farm in Northumberland to observe biosecurity measures, in this case neglecting to boil their pigswill and thereby cycling virus into the British herd (though the source of foot and mouth in the waste food also involved a border crossing – assuming foot and mouth was not already present in the UK, the disease may well have entered the country in illegal imports of meat from a country where foot and mouth was endemic) (Law and Mol, 2006). The resulting biocide or extermination

human disease. The exception is the H5N1 strain which had already crossed to humans in 1997 (in Hong Kong where there were 18 cases) and whose crossings have gradually increased in frequency and distribution in recent years. While individual cases have been tragic and caused large amounts of human suffering, outbreaks have remained limited in number. Notably, they have all occurred at the same time as H5N1 has been clinically present in nearby poultry flocks, and there have as yet been few if any clear cases of human-to-human transmission. All of which suggests that the current H5N1 virus is poorly adapted to its human host. For many, the more worrying danger is that H5N1 or another avian influenza virus will develop the capability to move quickly and effectively through human populations, as it undergoes reassortments (where avian and human viruses 'exchange' genetic material during a co-infection of a host pig or human) or through gradual adaptive mutations. While the latter may be detectable through a pattern of relatively small clusters of disease incidence and limited human-to-human transmission, reassortment may give rise to a rapid onset of a pandemic strain whose spread along existing networks could well prove so fast that current detection systems based on syndromic surveillance or even environmental monitoring followed by bioassay techniques would be too slow to signal a warning. The mutability and adaptability of viruses along with the complexity, intensity and density of animal–human and human–human interactions make for a complex political and policy environment.

How, then, is the bio being secured in such circumstances? One answer is to say that current securitisations are privileging a militarized, nation-state security. There is an escalation of border controls and purifications of territories, along with 'pre-emptive strikes' against violating states. In the 2001 UK foot and mouth outbreak there were certainly elements of nation-state securitization, especially as the disease progressed and ever more desperate attempts to develop strategies for eradicating the disease were devised (resulting in a militarization of the slaughter programme and a closure of the British countryside). Meanwhile, the reassertion of sovereignty is a particularly common refrain in work on US and international biosecurity. Mike Davis, for example, contrasts and directly links an overinvestment in a militarized response to possible bioterrorist activities with underinvestment in public health infrastructures in the US and poorly coordinated provision for future flu pandemics (including supply of anti-virals, broad-spectrum antibiotics, hospital beds and other infrastructures) (Davis, 2005). He makes parallel arguments concerning the international militarization of biosecurity at the same time as insecurities are being produced and intensified in the global slum city. The stockpiling of anti-viral drugs in the richer nation-states, coupled to their allocation to vital and military services should there be an outbreak, reproduces this geography of life chances. Meanwhile, a comparison of the responses to the threat of a flu pandemic compared to the lamentable performance of the pharmaceutical companies with respect to the AIDS epidemic on the African continent is testimony to the enaction of a particular mode and geography of securitization.

The war on bio-matters also involves, as Braun (2007) has shown, 'preemptive' strikes against human–nonhuman associations that are thought to be viral hot spots. The targets tend to be small mixed farms with poultry-raising facilities, pursued as a result of their reputed insecurity (their lack of regulation and their 'primitive' and bloody methods). However, as I will show, starting in Egypt, this selective targeting is controversial and subject to all manner of failure, and suggests that there is more to biosecurity than militarized modes of nation-state security.

After highly contagious forms of avian flu were detected in Egypt in the winter of 2005–6, the response was highly militarized. After ruling out vaccination of poultry on the basis that there was not enough vaccine stock available on the global market, the Egyptian government 'swung into action with a military-style cleansing operation. It ordered the culling of all backyard and rooftop poultry and banned live bird markets, where 80% of the nation's poultry is sold' (Grain, 2006: 1). In their place would be a different kind of agricultural practice and a different ecology. The Egyptian Prime Minister Ahmed Nazif described the new landscape thus: 'The world is moving towards big farms because they can be controlled under veterinarian supervision ... The time has come to get rid of the idea of breeding chickens on the roofs of houses' (ibid.: 1).

In place of the idea of rooftop agriculture would be the idea of industrial poultry production. This represented a hugely significant intervention into the political ecology of Egypt. Subsistence production in Egypt generally and small-scale animal husbandry in particular had taken on a new importance over the previous couple of decades. The story is complex but relates to shifts in rural land tenure and increases in rural landless labourers unable to make a living, to US government-led reductions in food subsidies from the 1970s onwards, to shifts in people's diets (from plant to animal products), to the resulting increases in food imports, and the privatization and deregulation of strategic food supplies (Mitchell, 2002). One estimate put the proportion of Cairo households that kept animals at 16 per cent rising to well over 25 per cent once the informal settlements and the former villages were included (Gertel and Samir, 2000). The keeping of birds – particularly chickens and ducks – was a way of life for people (especially women) in the low-income areas: a cheap way of adding otherwise expensive animal protein to the family diet (95 per cent of this farming was for home consumption). Home produced food was often perceived as being cleaner than bought-in meat and further acted as an economic buffer of sorts in times of increased hardship (ibid).

The targeting of small holders as enemies of the state of biosecurity is not only a matter for Egypt. World Health Organization reports convey the same message through images and text (see Figure 7.3). In one report, which aims 'to provide professionals with science-based answers to a number of common

**World Health
Organization**

QUESTIONS AND ANSWERS
ON
AVIAN INFLUENZA

A selection of frequently asked questions
on animals, food and water

Executive Version

Geneva, May 2006

Figure 7.3   Frontispiece to World Health Organization report on the
threat of Avian Influenza

questions about avian influenza' (WHO, 2006: 1), the aetiology, or narrative of disease causes and responsibilities, is made clear. It is wild birds that act as the reservoir for various strains of bird flu, and it is the mixing of wild birds with poorly controlled domestic bird populations that results in the spread of viruses into poultry. Finally, in this neat linear story, it is poorly regulated poultry handling practices where wild and domestic birds exist in close and visceral proximity to pigs and people that exacerbate the risk of highly infectious viruses crossing species barriers.

The WHO report reads as follows:

> In April 2005, the deaths of more than 6000 migratory birds, mostly bar-headed geese, due to the highly pathogenic H5N1 avian influenza virus, was reported at the Qinghai Lake nature reserve in central China. This event was very unusual and probably unprecedented. Scientific studies comparing viruses isolated from diseased birds in Africa, Europe and the Middle East have shown that they are almost identical to viruses recovered from dead birds at Qinghai Lake. Also, in countries affected more recently, diseased birds have all been found along the migratory routes of wild birds. While still poorly understood, it appears that the Qinghai Lake outbreak was the source of the westward spread of H5N1 avian influenza virus in birds in 2005–2006. Currently a total of at least 80 species of wild birds have been found to be infected by the H5N1 avian influenza virus. At least some migratory waterfowl have carried the H5N1 avian influenza virus in its highly pathogenic form, sometimes over long distances, and have infected poultry flocks in areas that lie along their migratory routes. (WHO, 2006: 2)

In the same report, the 'household', 'periurban' and 'backyard' enemies of biosecurity are emphasized when attention is turned to avian–human transmissions:

> Most human cases of H5N1 avian influenza have occurred in rural or periurban areas where many households keep small domestic poultry flocks. The H5N1 avian influenza virus is probably transmitted to humans through exposure during slaughter, defeathering, butchering and preparation of domestic poultry for cooking. (ibid.: 2)
>
> In backyard production settings, the system of marketing live birds and the practices of home slaughtering, defeathering and eviscerating, create opportunities for extensive human exposure to potentially contaminated parts of poultry. (ibid.: 4)

In the same vein, the United Nation's Food and Agriculture Organization (FAO) reversed its policy of promoting small-scale poultry farming (as a means to development, self-sufficiency and small business development). Instead it is the secure factory farms that should be encouraged (the only place for small open-air farms will be close to niche, i.e. wealthy Western, markets where presumably things can be closely monitored) (Grain, 2006: 3).

It is the absences as well as the presences in these reports that are important. Large, single-produce farms with tens of thousands of birds kept together in a handful of buildings are not implicated in avian flu. If there is any mention, they are the cure and not the cause of disease, despite their potential to provide ideal conditions for accelerations in viral mutation and reproduction. Similarly, there is no mention of the possibility that trade in birds and bird products could contribute to the spread of viral forms. Let me take production conditions and movements in turn.

For Mike Davis, the vilification of specific forms of agriculture plays directly into the hands of organizations which would be set to gain from greater degrees of centralized control on the animal-rearing business. The industry argument runs that organizations like Tyson Foods, which kills 2.2 billion chickens annually (Davis, 2005: 83), would be in a better position than small producers in terms of conforming to biosecurity standards (like, for example, specific modes of worker apparel, use of anti-viral drugs in livestock, standardized farm buildings, and so on). Yet it may well be that these types of production facilities are the problem not the solution. The density of populations, the systematic use of antibiotics and the resultant production of resistant strains, the long-distance movements of livestock, livestock products and dead stock, the uniformity of the gene pool, the known stresses that industrially raised animals suffer – all these ecologies of production contribute to an increasing prevalence of species-specific and zoonotic viruses, and the probabilities of further viral mutation.

In terms of domestic bird and bird product movements, an alternative science of connections is possible, turning the WHO's aetiology on its head. What if avian flu is predominantly a disease of wild birds caught from commercial poultry and not the other way around (Grain, 2006)? Qinghai Lake in northern China, for example, is surrounded by intensive poultry factories, and the manure generated by the birds is used as feed on the many fish farms that exist on the lake. It is also used as fertilizer on fields around the lake (Blythman, 2006). It is well known that bird flu viruses can survive in faeces for over a month in cold temperatures (WHO, 2006: 4). Whether or not this spreading of virus can then infect wild birds is arguable, but the possibility is there. Likewise, the international mass movements of birds and bird products from factories may just as easily be implicated in a disease network as migrating birds. Indeed, there are arguments that the viral pathways match export rather than seasonal migration routes (Blythman, 2006; Grain, 2006). It may well be that the focus of biosecurity should not be the wild inputs into small farms but the fermentations on large commercial holdings and the wild spatial array of markets that exist downstream of the poultry industry.

So, while there are clearly ripe conditions for species crossings in mixed farm settings, the conglomerate industrial producers are hardly innocent players in pandemic ecologies. The broader point here is that the 'modernization' and industrialization of agriculture are being sutured to a securitization of sovereign states (notably those of the rich North). The concurrent production of global insecurities is evident in the realization that such modernization inevitably, it seems, produces as many if not more opportunities for the co-production of viruses with the potential to produce pandemics. So much, again, for modernity. Purity breeds impurity, and the denial of the latter adds fuel to the modern's evangelical claims (Latour, 1993).

Multiple networks

If one of the arguments that arise from these examples is the impossibility of purity, a second is that modes of securing are themselves multiple, and prac-tised in many different places, involving many different things, not all of which will lend themselves fully to the programme of action. That is, rather than view security as logics, carried out to the letter and with various in-built bias and social interests, we might engage with biosecurity as practised in a variety of ways, the results of which are more open. In renaming the logics of security as modes of securing, the aim is to demonstrate not the march of a single (state or capital) logic that should be resisted or demonstrated to be full of already existing (social) contradictions, but the multiple and heterogeneous ways in which securities are enacted.

The slaughter of livestock in the UK and Egypt along with farm isolations, removal of livelihood, and wholesale changes to political ecologies of live-stock suggest, perhaps, a militarization of formerly civilian spaces, and a resulting curtailment of civil liberties. Indeed, there is a common critical move in social science to talk of a militarization of geopolitics (see, for example, Graham, 2004). But such securitizations are also a matter of surveillance and of self-regulation, in ways that are not simply about territory and securing boundaries but about the policing of populations (chickens and people) and the extension of some liberties (e.g. trade and large corporations) over others (animals, smallholders). In addition, they are also far from being complete. Indeed, in Egypt and the UK and in practice things did not work out so cleanly. In the UK non-compliance with ordered culls is well known. In Egypt, rural householders in particular hid their birds in fields and refused to let apparently healthy birds join the cull. Sub-clinical birds were rushed to markets, a practice that was encouraged given that state compensation was lower than market prices. When the disease struck, apparently unaffected birds were killed in order to 'rescue them' (as villages called it) from the dis-ease (Slackman, 2006). Many kept their ducks as these seemed to stay healthy (but ducks are good at staying healthy even while they host avian flu). The cull itself has been criticized, as military methods combined with the secretion of live household birds to keep them safe and the dumping of diseased birds in canals and waterways may have risked a greater spread of the disease. These practical difficulties led to a concession in rural areas that flocks could be kept as long as they were caged and healthy, and was accompanied by a vaccination/surveillance programme.

As I have noted there are different types of security (Collier and Lakoff, 2006a), or better multiple practices that co-exist with one another, pulling and shaping the worlds they enact in different ways and with different effects. The question may then be, how do these modes of securing get along together? Does one replace the other? Does one dictate terms to the other? Do they work side by side? Do they interfere with one another or even contradict one another? The answers will be complex, and dependent on empirical situations, but in the Egyptian avian flu case there are at least two things to say.

First of all, there are some fairly clear contradictions between modes of securing. Disciplining poultry production through its agglomeration, for example, may not be simply a subservient from of securing enacted in the service of nation-state security. Immediately after ordering the cull, the Egyptian government had to ease import restrictions on frozen meat to make up for the domestic shortfalls. It also started to implement a re-stocking of its commercial farms with chicks from the US and Europe. So, in the short term at least, it is unclear how the shift from small producers to commercial factories aids national security in Egypt or elsewhere. Likewise, with an expansion of intensive poultry farming, there is every chance the strategy will backfire in other ways. In short, the relationship between modes of securing is more complicated than one of hierarchy or one where there is a dominant mode of ordering to which all other forms of ordering follow suit. Second, beyond hierarchy and contradiction, other relations are possible. In the cull of animals in both the UK and Egypt, there were episodes of confusion, concession, adaptation and accommodation. In short, programmes fail for a host of contingencies.

The aim of such analyses is to demonstrate that rather than militarization existing as a single or even singular logic which we should critique (and thereby risk making it sound more pervasive than it might well be), it is a set of practices that are limited in extent, far from foundational and thereby open to disruptive elements. In emphasizing that there is more than one mode in operation and that all modes are themselves imperfect and practised, biosecurity will therefore involve the imperfect juxtaposition of a number of matters, spaces, temporalities, drawing relations with and between farmers, consumers, blood, abattoirs, international trade, markets, tourists, governments, the UN, large food corporations, to name just a few. All these matters, relations and topologies are bound to be productive, to throw up surprises, to, as Deleuze puts it, churn 'up matter and functions in a way that is likely to create change' (1999: 35; see also Thacker, 2005a). In many ways, this relates to the points made in previous chapters on multiplicity and openness (Massey, 2005). And it relates to one of the main arguments of this book – that attempts to organize matters will necessarily involve, enfold, matters that are loosely connected to the strategies and schemes thought to be in operation. In other words, there will be spaces for nature. So a feature of the complex practices of biosecurity is that they draw in and mix up all manner of things, materialities, socialities, and so on. The result is not always predictable (indeed, it is far from predictable).

Thacker's (2005a; 2005b) work is particularly helpful here as he traces not so much the imposition of a specific security logic onto a new and nebulous enemy (the critical trope suggesting that the old logics and devices of nation-state security, with their volumetric topological spaces, are incompatible with new threats, including biological and terrorist networks, is a common one), but a more open-ended and topologically charged process of what the US defense industry refers to as netwars. 'This is a situation of what we can call networks fighting networks, in which one type of network is positioned against another, and the opposing topologies made to confront each other's respective strengths, robustness, and flexibilities' (Thacker, 2005a: 8). One network in these netwars would be the disease, attempting to produce a distributed mode of existence, one that is self-organizing, mutable and in process. Another would be the surveillance and regulatory frameworks that are distributed in ways that can detect, warn and curtail a disease network. These and other networks will call on shared elements in different ways. An international traveller, for example, will be called upon as viral host and as a regulated and profiled citizen.

The famous case here has become the relative success of the WHO's efforts to combat the SARS (Severe Acute Respiratory Syndrome) coronavirus. The 2003 outbreak spread rapidly through air travel networks, turning up in global cities including Beijing, Hong Kong, Singapore, and Toronto in rapid succession (it's an often repeated point that as travel times have become shorter than incubation periods, then the disease can network over long distances before it is identified through clinical symptoms). The mobilization against the disease involved preparedness and response activities (including the interception and quarantine of suspected carriers on international flights) which folded together 'not just technology [in the form of computer networks and e-conferencing facilities], but also negotiations among WHO officials with governments and hospitals, from Toronto to Beijing. All these processes of information exchange and communications constituted part of the WHO's counter-epidemic network' (Thacker, 2005a: 8).

Yet as Thacker goes on to suggest, the topological realization of netwars is possibly not radical enough – for in the current design of biosurveillance and disease surveillance networks (DSNs), there remains a logic of control and instrumentalism, one that underestimates the nonhuman and thereby indeterminate characteristics of networks. Indeed, viral networks can develop orderings that are highly resistant to top-down, centralized forms of control. Surveillant systems that are too rigid (like electro-mechanical devices that cannot respond to novel strains, or are in the wrong locations), decision-making algorithms used to record positive disease data that leave no space for a multiplicity of symptoms and even the widespread culling of animals in order to produce a disease-free zone – all may mark the wrong kinds of relation between networks and between modes of securing (see Box 7.1).

Box 7.1   Multiple securities: comparing surveillant technologies for military and public health securities

Electro-mechanical devices used for environmental sensing are being placed at key locations in US cities and increasingly elsewhere. Derived from military applications, they are designed to give an early warning of the presence of a biological agent that could be harmful to people. They make use of a variety of technologies designed to compare the content of an air sample with the identities of known pathogens. This can involve the use of polymerase chain reactions (PCR) to produce sufficient copies of DNA present in a sample to carry out tests in order to identify a microbe. Two limitations of the technology are, first, the production of false positives and false negatives, and, second, the limited range of pathogens that they can recognize.

The derivation of these devices from 'ideal' (and possibly rather outdated) military situations where they are designed to warn of chemical attack gives clues as to their capacities and limitations. In warfare, biological and chemical agents are often known forms, predictable, visible and act quickly. The battle fronts and therefore directions of threat are known, making preparation more straightforward. There are specific entities that will appear over the battle horizon on fairly well-known days (the weather conditions are normally right for this kind of attack, i.e. dry, consistent wind direction blowing towards 'targets', etc.). Sensors can be arranged to face a known threat. Detection here works to a science of known agents and known enemies. It is underpinned by a secure form of knowing.

In the case of civilian bio-threats, agents are less predictable, both in form and in behaviour. Viruses may contain mutations or reassortments that make them difficult to detect using PCR. They are likely to be odourless and invisible. Clinical symptoms may take weeks to appear after onset or attack. There is no obvious front (though various networks including urban infrastructures, transport systems, postal systems, sports stadia and corporate headquarters figure highly in risk strategies). It may be the case that the most likely source is not from an enemy at all, but is generated by accident and/or through human–animal crossings.

(Continued)

(Continued)

This complexity suggests that 'civilian' biosecurity surveillance needs to occur on many sites, using many different forms and large volumes of sense data. Other forms of surveillance include prodromic data (looking for immunological markers to detect a disease prior to clinical symptoms developing), syndromic surveillance data (based on monitoring data from hospitals and doctors' surgeries to detect abnormal admissions), pharmaceutical sales data (again to detect abnormal increases in painkiller sales, for example), the use of sentinel organisms (nonhumans known to develop symptoms in advance of humans) and veterinary surveillance data. All of these methods have their problems, not least the handling of false positives and the risk of false negatives, but many at least have the potential to adapt to novel biological agents. The main implication is that 'civilian' biosecurity is multiple, it involves many different places, unknown and possibly unknowable biological agents, many different organisms, populations, machines, data, computer networks, aeroplanes, farms – the list is most probably endless. Knowledge is less sure of itself. Instead of dealing in knowns and risks (probabilities of attack), it needs to handle unknowns and indeterminacies.

The challenge of establishing sovereignty within a network becomes a necessary paradox, where the need for control is also the need for an absence of control (Thacker, 2005a: 13) (see the previous chapter for an illustration with respect to BSE). Echoing the conclusion of Chapter 6, it may just be that the looseness of a network, its adaptabilities and accommodations, is what gives it strength. And perhaps the war metaphor is not therefore quite right – for it may be the conditional hospitalities that make for a prepared society. So, in suggesting that biosecurities are multiple, and in starting to develop an understanding of how those multiple practices can co-exist, we need more than contradictions and conflicts to guide our thinking. It may well be that in addition there is a need to think about how to find other ways of collectively living with disease than imagining that we are perpetually at war with it.

Conclusion

There have been two arguments developed in this chapter. The first is now possibly familiar. It is that attempts to order will provide conditions for disorder. There is a relation here of negation. Gardens make weeds (Callon and Law, 2005 and Chapter 10), clean herds make dirty herds. They also make topologies, both of regions which emphasize separations (clean or dirty) and networks which emphasize connections and interrelations. Ever more desperate attempts to maintain distances between regions, through, for example, mass culls and or through centralization and self-regulation of farming methods, seem to make connections or overflows more rather than less likely. The second argument takes things forward a little. It has been to suggest that these attempts to order are themselves already complex and heterogeneous practices that will relate with many other practices. The result is that biosecurities are multiple, and can pull and push all sorts of disease objects in different directions, with sometimes surprising results. The relations within and between those practices will also be highly differentiated. Sometimes conflict and military-style activities come to the fore, but they are never the only practice or goings-on that are making matters. Other practices exist side by side with one another, are indifferent to one another, annoy but then live with one another. In later chapters I will tease out these and more possibilities for living with difference. Here the main aim has been to suggest that biosecurity may be best thought of as a multiplicity of things and relations that are in process and are unlikely to form a settled state of affairs.

Background reading

Mike Davis' timely book *The Monster at Our Door* (2005) is a must read for anyone interested in questions of biosecurity. For an insightful look at Foucault's bio-politics and its extension to biosecurity, see the essay by Braun (2007).

Further reading

John Law's interventions (some with Annemarie Mol) into the foot and mouth epidemic provide a clear insight into the difficulties of tracing the politics of a disease (Law, 2006a: 2006b; Law and Mol, 2006). Eugene Thacker's (2005a) work comes at similar problems from a more Foucauldian angle.

Conserving natures

Hybrids, conformations, relations – all these help us to discuss natures as matters that are not simply natural. We've loosened the 'nature as independent state' metaphor. This has been all very well for molecular sciences, for agriculture and for biosecurity – but what about that quintessentially romantic and nature-independent activity 'nature conservation'? Surely this isn't well served by our complex geographies of nature? This chapter explores this problem, and in doing so adds further detail to the difference of nature issue that was raised in Chapter 5.

The setting for this chapter (and for Chapter 10) is predominantly urban. This might be thought of as the last place one would go to in order to conserve natures. Indeed, one of the more enduring elements of anti-urban, Euro-American environmental thinking is a firm separation between cities and nature. Cities are not where we go to find nature. The 'nature as independent state' story works to sever connections between cities and natures. Futhermore, it's not just environmentalism that is guilty here, the obsessions of social and cultural theory with cities, as highpoints of cultural intensity (Massey, 2005), reproduce the notion that there is nothing much of nature in the city. This chapter aims to unsettle this dichotomy by demonstrating the complex relations between cities and natures and between humans and non-humans. The chapter moves towards considering what kinds of relations are needed for conservation, arguing for a double recognition of both association and dissociation, proximity and distance, joining and separation in any attempt to build a conservation politics.

Conserving nature in theory and practice

The anthropologist Marilyn Strathern suggests that 'The divides of modern people's thinking do not correspond to the methods they actually employ' (1996: 522). As I will demonstrate, nature conservation's divisions do indeed look less formidable when we follow its practices rather than theories (see also Latour, 2004b).

Implicit in many of nature conservation's manifestations are the following:

1. There is a rationale to retain something of a pre-existing state of nature.
2. Nature in other words is pre-constituted and conservation comes after nature.

3. In order to save its object from being extinguished, nature has to be present (here and now).
4. And the present nature has to represent something of value (a species, a genotype, a habitat, a land management regime).
5. The final act, as Geoffrey Bowker has put it, is 'to render the present eternal' (Bowker, 2004: 112).

In practice, things are often much more difficult. First of all, it is not always helpful to start with a fixed version of nature, or with a firmly held sense of what the issues are. This is partly a matter of theoretical clarity. We have already established that nature can't be thought as somehow evacuated culture, the blank space left when values, politics and so on have gone. Indeed, we have unsettled the equally damaging notion that nature is the dead-end 'same', to be compared either favourably or unfavourably to cultural diversity. Rather, we have started to pluralize and animate natures, not just as others to all those cultures, but as multiple, differentiated spaces that are neither underwritten by nor always clearly distinct from cultures. Meanwhile, and in addition to these theoretical advances which are, to be fair, often less important to conservation practitioners, the natures with which people work are often far from being known, understood, predictable or matters only of control. The object of nature is not easily represented, managed or brought into politics. This chapter provides examples where, rather than being pre-set matters which only need better management, natures are in the making. Any engagement with nature conservation must, then, engage with this reconstitution of natures.

Second, the assumption of presence is often difficult in practice. There are two elements to this problem. On the one hand, in conventional terms, living assemblages (organisms, habitats, landscapes) have to be present, or made present (especially if we are talking restoration) if they are to be conserved. On the other, the embodied present thing also often needs to be *representative* of other things in order to gain conservation value. This is particularly the case in the science of animal and plant conservation, although it can also be invoked in other assemblages like specific habitats (in conservation strategies, quotas are often set for habitat types such as heath, grassland, deciduous woodland, and so on). The organism or other speaks *as* a specific embodiment of a species kind. So a small, black bird, with a flash of red on its tail, and a distinctive, if unremarkable, song, perching on a crumbling brick wall in a car park right at the centre of Britain's second city (Birmingham) is not only present in the city, it also represents a species, in this case black redstarts (*Phoenicurus ochruros*), Britain's rarest bird. Similarly, three kilometres south of the city centre, a small mammal known as a water vole (*Arvicola terrestris*) manages to evade the various hazards of inner city life. Both species are in danger of extinction, of becoming absent, the mark of failure for nature conservation. And yet, their endangered status is only part of the problem. For in practice their presence in the city is far from being self-evident. It can be extremely difficult to locate black redstarts and water voles, and there are even arguments that some of these urban inhabitants fail to represent their species satisfactorily (they are too urban). So in practice, not only are natures in process, presence starts to become an unreliable platform on which to build a conservation argument.

One way of comparing conservation in theory with conservation in practice is to denounce conservation practice for failing to live up to its own claims. This is one area where natural and social scientists develop similar forms of critique. The targets are familiar. Too many mega-fauna and too many charismatic and exotic creatures claim ecological priority. Individual species dominate programmes rather than broader assemblages or ecologies. Rarity is given too much attention. Stasis is favoured over process. Funding skews action. There's a Hollywoodization of conservation, and a resultant privatization of wildlife parks (Adams, 2004). Even before private parks, national parks were never natural. Meanwhile, the expertocracies of first world conservation science stamp all over indigenous, lay and popular concerns (see Chapter 1).

While many of the issues raised are of pressing concern, it should not be assumed that the *facts* of conservation, the real issues, pre-exist action and are thereby invariably polluted by human values, politics, cultures, ways of seeing, and so on. On closer inspection, the facts are far from settled for at least two reasons. First, taking facts as the only guides for action would lead to rather outlandish outcomes. In the extreme, 'real conservation' in the form of maximizing biodiversity potential would, assuming a molecular clock model of genetic change (a big assumption), result in it really only making sense to conserve the earliest diverging lineages of bacteria (Williams et al., 1994). Second, and more importantly, in practice facts are not so smooth or clear. Indeed, I want to argue that as with other experimental endeavours, facts emerge through practice, in the fields of activity, and are not prior to those practices (see also Chapter 3). The implication is that if facts are made in practice, then they are far more precarious than we might have been prone to believe (Mol, 2002; Rheinberger, 1997). Thus, the objects of conservation (like species and habitats) are not fully formed or always fully present, but in the process of being made present.

In denying that the facts of the matter are available prior to the practice of conservation, I am certainly not wishing to denude nature conservation of things, nonhumans, animals, plants, genes, and so on. While the facts of the matter are not clear bases for politics, nor are the matters of facts irrelevant. Far from it. It is the very plenitude of nature conservation's objects that make for a world where facts are in process rather than clear-cut issues. In short, I'm interested here in the enaction or making of conservation and natures, a making that is always a more than human endeavour (Whatmore, 2004).

In the next section I look in some detail at the fraught practices of making something present. Later I turn to the question of what exactly those present represent. Finally, I reflect further on the spatialities of conservation politics.

Making things present

Making things present is just that – a making. It is not a revealing and it has a history. It's now common, for example, to say that the practice of presencing is somewhat different than was the case in the past. At the time of the travel writer's W.H. Hudson's fascinating 1890s study of *Birds in London*, 'serious bird watchers ... were almost obliged to shoot any strange bird in order to identify it. Nobody believed a purely sight record, because it was thought that most birds, and certainly rare ones, could not be identified with any confidence except in the hand' (Fitter, 1969: iv.) Today, outside certain fields like entomology, the taking of dead specimens is a rarity and the private collector is almost extinct. So presence in some quarters at least is not quite what it was. Rather than a presence of absence enveloped in the dead specimen, presence is produced through all manner of practices. In order to make the argument I will look in some detail at two field examples, set in a British city (Birmingham).

Making present 1: urban water voles

Let me set the scene by describing, twice, a piece of land three kilometres from the centre of Britain's second city. It is a large site, nearly 50 hectares in total, most of which is former industrial and allotment land. One way of characterizing the site would be to say that it is a mosaic of habitats (from spoil tip, supporting a range of thermophillic plants and insects, to wetland assemblages in hollows formed by quarrying). The site is used by a variety of species which reside or pass through, making use of the corridors and patchworks of wild spaces in this part of the city (the site is bordered and crossed by railway lines, canals and a stream, all of which can act as passages through the city for plants and animals. There are also domestic gardens, parks and school playing fields in the vicinity, also aiding plant and animal dispersal – see Figures 8.1 and 8.2).

Another way of describing the site would be to note that this part of the city has been designated as a high technology corridor. The logic here is that the site might become somewhere that capital projects and market relations can more easily settle, or pass through. Indeed, a significant part of the site is earmarked for a new private finance initiative (PFI) superhospital, another section for a supermarket, developments that will require a relief road to be constructed. Once in place, the accessibility provided by the road will enable the remaining parcels of land on the site to be re-developed.

Figure 8.1   The Bournbrook, two miles from Birmingham's city centre

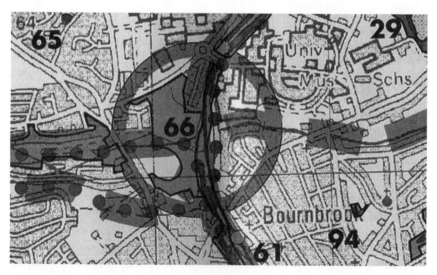

Figure 8.2   Detail from the City of Birmingham's Conservation Strategy depicting the case study site (numbered 66), designated as a Site of Local Importance for Nature Conservation

Note: The thick solid lines indicate 'key wildlife corridors', the dashed lines 'wildlife corridors', the circle indicates a node or confluence of more than four corridors, the dotted lines indicate linear open space – the Bournbrook stream is the faint line running across the middle of the map.

These two corridor visions are at least in part in conflict – one seems to prohibit the other. There's always more than one way of describing a place – perhaps there are less than many ways – or to put this another way, the possibilities are not infinite (especially when it comes to the fraught processes of ontological politics). So how to proceed and proceed well when it comes to deciding which vision should be carried into the future?

One way would be to look to nature conservation in order to assess the nature on the site and then find a means to render that nature eternal. This would make life politically more straightforward – nature conservationists might be the external arbiters, deciding whether the site is worthy of salvation from the developers. There are problems with this view of course. One of these is that it is difficult to get an agreement on what is actually present on site. Let me evoke some of this difficulty by relaying some of the processes involved in learning how to read the landscape for water voles.

The site is a promising habitat for water voles, with its stream, wetland, willow trees and ground vegetation. There are at least two further causes for optimism. First, the stream is at the upper end of the River Trent catchment, so if the water vole is in retreat we might expect animals in this part of the river system. Second, the water vole's deadliest predator, the American Mink, hasn't yet been recorded in Birmingham, so local ecologists think that the urban area might be good refuge for an animal that is close to extinction elsewhere. Against these positive aspects of the site, there are many matters that militate against water voles. The urban river is prone to rapid changes in level as it sometimes acts as a storm run-off channel. Bank-side burrows can therefore suffer rapid flooding. Related to this there is water pollution, especially after storm events which can produce peaks in contaminated run-off reaching water courses. There are also lots of predators around, including brown rats, domestic cats and birds of prey.

In the recent past and in England and Wales, water voles have been numerous and so have been relatively easy to find (although they are seldom obvious to the untrained eye, often being mistaken for brown rats). They have a 4-hour metabolic cycle of foraging and sleeping, and are slightly more active in daylight hours, so spotting water voles may be a matter of waiting for them to emerge from their burrows. However, with low numbers and in areas where water voles are unlikely to be as bold (where there is frequent predation and human traffic, for example), they may adopt a less conspicuous habit. In these cases, learning to make water voles present involves being trained to read the landscape for other traces – the footprints and other marks that water voles leave or make as they move around, forage and mark territory. I will call these traces water vole writing. Writing is a risky term. By using it I aim to unsettle the sense that nonhumans are always merely written up. In ceding at least some of the writing action to water voles and other characters in this story I can hint at their contributions to what Deleuze and Guattari (1988: 141) call the material processes of writing.

On the first water vole training day, the group starts by looking for footprints, gazing about the ground with little confidence or direction. Our first problem is finding anything that can be described as a footprint. A few days of rain and thick summer vegetation mean that signs are difficult to locate. When a series of footprints is spotted across a mud-covered concrete support we are introduced to a second problem. Prints are far from self-evident – to the unversed this might have been any small creature, even a bird. So, out comes the field guide, which, disappointingly, is far from being a definitive guide to the field (see also Law and Lynch, 1990). We learn that water vole footprints are similar to those of the brown rat – slightly smaller and more star-shaped with the main adjudicator seeming to be the angle set by the first and fourth toes (see Figure 8.3). But when you add in the complexities of substrate, the variety of animal sizes that exist from juvenile to adult, the different speeds of movement, varieties of slope, and that water voles tend to cover their tracks (by treading on their foreprints with their back paws), then things

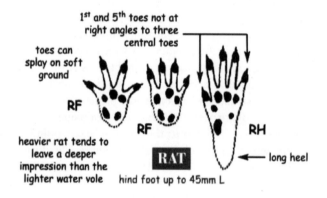

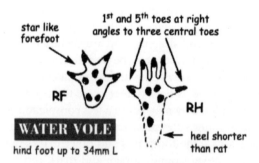

Figure 8.3   Field guides to spotting wildlife – differences on paper serve to diagram, or develop a learning to be affected, that allows for differences to be detected in the field

become difficult. While we deliberate over this series of contingencies, one of the group adds that the vulnerable water vole tends to avoid such open areas. Rats, on the other hand, are bolder. The consensus is that the traces we had in front of us were not written by a water vole, and were most likely to have been a rat. Spotting prints and distinguishing water vole writing from brown rat writing required fine tuning our observations, comparing field prints to figures in a book, allowing for all manner of contingencies and weighing up corroborating or mitigating evidence.

After failing to find any more tracks, the group moves from the stream to the wetland, where, in the absence of any exposed track-bearing mud, we look for the 'tell-tale' latrines. Water voles, it is told, are often fastidious creatures, who have designated latrines, trampling down their earlier efforts before adding fresh material. Rats, on the other hand, tend to go anywhere and everywhere. These latrines are relatively easy to spot, to the trained eye. And if the organization of faeces isn't clearly differentiated in the field, we're told that rats' faeces tend to smell differently from those produced by water voles. Again, though, we fail to find any evidence of water vole. Perhaps, after all, they don't inhabit our site. The consultants employed by the developers who want to build on this ground say as much (Babtie, 2001).

The next Spring, in conditions that are more favourable for reading water vole writing (it had been dry for a number of days and bankside vegetation was not yet abundant), we visit the site again. This time faeces figure highly. The size, shape, organization and smell of faeces enable us to identify water vole presence on the site and upstream. We become adept at spotting the small latrines on the mud banks. Eyes trained, we learn to handle the faeces, breaking them apart to sense the animal's diet (water voles are vegetarians, rats omnivorous). We become noses too, although that is possibly too grand a term. Unlike the wine and perfume noses that Latour (2004a) and Thrift (2003) refer to, our bodies and noses seem almost 'tuned' or trained already to the smell of rat faeces. Omnivores produce far more pungent faeces than the rush-eating water voles.

We learn too about water vole lawns and runs – neat grazing areas and well-trodden paths. We learn that the angle at which a rush has been severed can be indicative of water vole actions. We learn to add field observation to field observation – to discuss, to collaborate, to corroborate, to build a sense of water vole inhabitation.

Are we representing the water vole? On the one hand, yes, this is a process of recording presence (or absence) and being able to re-present this elsewhere. On the other hand, something else was going on here. In practice, the pictures and written texts are woven together with the traces, tracks and mammals to form a complex of writings. Our eyes (and to a lesser extent our noses) were being trained to recognize distinctions that were formerly invisible to us. The pictures, field signs and conversations were changing the way we sensed and, as we will see later, the way water voles made sense. I'm not sure that representation is the best term to grasp this complex of activities and interactions. An alternative would be diagramming, which conveys a sense of 'writing around' water voles. Field guides, for example, write around rather than write up, once and for all, their object. Learning water vole writing involves rapid movements between texts, descriptions, field signs, conversations, comparisons, finding similarities, explaining differences and so on. It involves a view from whereabouts (see Chapter 1). To be a good reader required a form of expertise that could combine multiple indications of presence, a looser kind of sense, a knowing around water voles, diagnostics and a diagramming.

Those of us previously unversed in water vole writing started to look at the landscape rather differently. As Latour (2004a) might say, we had started to learn to be affected. We were bodies in process, gaining ways of looking, a new set of eyes (or newly conditioned retina), slightly more wary nose, a different sensibility. The role of texts is important to grasp here. The field guides as textual inscriptions of footprints and other characteristics were not equivalent to footprints in the field. But nor were they either rich or poor depictions. They were, instead, sensitizing devices, diagrams, that made water voles more rather than less real for those who started to use them. The texts certainly helped us to spatialize the landscape, to read this part of Birmingham, noting species' difference and their possible presence or absence in the landscape. However, these spatializations were far from being textually determined. Instead, the texts acted to alert us to rather than define water vole spaces.

Instead of faithful representations, we're starting to talk about a creative address, an address that engages with the world rather than imagines that it can simply present it. I will have more to say about some of the consequences for conservation of this non-representational element of survey work in a moment, but one point to emphasize is that making water voles present is not as straightforward as it sounds. Work is needed to make water voles present and, importantly, water voles are not simply made present but are also involved in the process of writing traces. The result of this coming together of traces can produce surprises which challenge the notion that field surveys are only matters of recording that which is already present. Instead, they are also matters of recognizing *what* is present.

If we were to say that we are simply generating faithful representations of water voles, then it assumes that water voles, as water voles, are there to be represented. They are, effectively, self-identical species that can be counted – scientifically and, with the necessary politicization of ecology, democratically. Note that water voles are expected to look the same at the start of this process as they did at the end. This dream of the one-to-one map, and its green democratic equivalent, one vole, one vote, can be spotted in practices that are being performed in the name of biodiversity and nature conservation

(Bowker, 2000). Ideal forms (often but not only species) are turned into manageable and manipulable data.

In practice, though, things are a bit muddier than this dream of faithful mimesis. And it is through, rather than in spite of, this mud that other possibilities for action can be seeded. I can begin to demonstrate how by following the practices of the ecologists involved in this site a little further. Ecologists at the Birmingham and Black Country Wildlife Trust have been experimenting with other ways of printing and reading water vole writing. They are using 'tracking tunnels', in which a sponge soaked in food colouring is overlain with paper to record footprints. This may seem an unremarkable extension of the technologies that I have already described for sensing water voles. But, something interesting may have been produced. The ecologists identified water vole activity on our site using their tunnels. But, what was more surprising, they also suspected that the same tunnels were visited by brown rats. Footprints of both animals seemed, to the ecologists, to appear on the same pieces of blotting paper in their recording device. The received wisdom is that water voles and brown rats do not co-habit. Standard (ruralist) ecology presumes that the presence of brown rats results in rapid water vole population decline, through rat predation, competition for food and territory. Here something different seemed to be happening. A question raised for the ecologists was whether this indication of co-existence had any significance in terms of the urban ecologies of water voles. Could it be that the urban water voles are actually different from their rural counterparts in terms of their ecological relations?

For these scientists, learning to be affected is rarely a matter of trusting one piece of evidence. Indeed, the writings in ink, just as much if not more than the field signs, are far from self-evident inscriptions. So again, it's back to the process of generating more references. As mentioned earlier, the organization and smell of faeces can stand for water voles and/or rats. On several ledges, close to burrows that could be inhabited by brown rats or water voles, the ecologists have recorded water vole latrines with rat droppings close by or on top of them. This overwriting has been going on now for two years. It seems more than likely then that, for a moment at least, rats and water voles are co-habiting on the banks of our urban stream. Learning water vole and rat writing has enabled something new to be said. It's tentative to be sure (although the ecologists are adamant that co-habitation of some form is occurring).

Beyond marking presence or absence, we're starting to question what water voles and other urban cosmopolites *do* in the city. Water voles have started to change – they are no longer quite the same species with which we started. Sensitivity to minor differences may or may not prove fruitful. But it seems that, as with other scientific endeavours, this openness to difference, which is borne out of a looser kind of sense, a knowing around rather than knowledge of, is a vital means to allow for nonhuman knowledgeabilities. This is, for some, the mark of good science (see Latour, 2004a; Stengers, 1997). Indeed, it is the ability to listen to the vagueness of the epistemic thing (Rheinberger, 1997), or the putative object of inquiry, which is a condition of possibility for new knowledge (see earlier chapters). It is, as another way of saying this, the ability to address nonhumans as colleagues in the process of producing knowledge that makes new knowledge possible. It's a form of address, then, that treats people and water voles (in this case) as fellow subjects rather than as pre-formed objects. For Stengers, this ability to listen attentively is a way or means of putting knowledge at risk and allowing others, of all shapes and sizes, to make a difference to the process of knowing (Paulson, 2001; Stengers, 1997).

For the ecologists, this is an interesting moment in water vole studies. It challenges the universalism of water vole ecology. More than this, the urban stream and habitat are becoming more interesting than pale imitations of the rural idyll. There is what some ecologists call a recombinant ecology here (although I'm slightly extending the remit of the term in this instance), which is so much more than a relict ecology or a restored ecology (Barker, 2000). Its times and spaces are quite different from the *representative* ecologies that often aim to mimic a distant memory. Attention is turned to unfinished urban collectives. The urban topology and lifelines (Rose, 1998) that this site's space-times engender become more of an issue. How does this site connect to the rest of the river network? What about its own mosaic of habitats? How do these enable water voles to survive despite the presence of predators? What are the time-spaces involved? Have the species been cohabiting for a period of years or is this a moment in the readjustment of populations? The questions start to proliferate.

There's a double movement here – of people and things. It's important to be clear that I am not saying simply that water voles and the site are 'in reality' different from old, less accurate representations. Rather, in the reality that is being so imaginatively fabricated with the references that are being co-produced, people, water voles and the site are starting to address each other in a different fashion. New possibilities are opening up. The wild things in Birmingham are suggestive of a more interesting political science than the marking of species presence or absence, a political science that doesn't take so many shortcuts. It's no longer a matter of saying once and for all that water voles are present or absent, as if this statement could bypass politics and therefore declare peace. Rather, there is a more fraught politics on the table, one that proposes new possibilities and therefore requires a very different kind of peace settlement (Latour, 2003). I will return to the water voles in a later section, but it will be useful at this point to look at another case of making present in order to underline the tension between the openness of conservation's field practices and the potential closures of conservation's legal apparatus.

Making present 2: black redstarts

Here is another example that will be useful to us in developing a new diagram for nature conservation. It starts in the same city and involves black redstarts, reportedly Britain's rarest bird. There are around 100 in Britain. Five million are thought to exist on mainland Europe – making their conservation in Britain ripe for ideology critique. It could be said that conservation is so bound up with nation-states that it fails to see the British black redstarts as merely outlying individuals at the edge of a species' range. A species, moreover, that is far from being at risk of extinction and that is a relatively recent arrival to the British Isles (first reports of sightings were in the 1920s). While such a quick denunciation is tempting, I want to suspend such judgements, as there is another politics to pursue.

In Britain, black redstarts are predominantly found in or close to city centres, with Birmingham and London the most frequented locations. They migrate here from mainland Europe (although some have been known to over-winter in the city) and, while in Britain, are at the edge of their normal range. Numbers are small. In Birmingham there were thought to be 15 breeding pairs in the 1980s (boom time, in relative terms, for black redstarts). Thatcher's Britain was good for black redstarts, as were other periods of urban decline. A partial black redstart history suggests the bird has had three periods of relative success in the UK. Black redstart populations have been at their highest levels in the late 1920s and the early 1930s, in the 1940s and then in the 1980s. So the bird has enjoyed relative success at times of human economic and urban decline or when there's a human war (black redstarts take advantage of bomber ecologies as Davis (2002) calls them, ecological assemblages that inhabit disused buildings and bombsites).

The urban renaissance in the UK hasn't been good for black redstarts. Since the 1980s, ancestral breeding sites around places like Gas Street Basin in the centre of Birmingham have been turned from canal-side industrial dereliction to landscapes for human consumption. The loss of suitable habitat has very possibly contributed to a decline in the Birmingham black redstart population, which is now thought to be something like two to four breeding pairs (still 5–10 per cent of the UK population).

National rarity endows black redstarts with such ecological importance in the traditional value system of established conservation that it receives as much legal protection as is currently attainable in the UK. There is a robust legal framework (including designation as a fully protected species on Schedule I of the Wildlife and Countryside Act 1981 (as amended)). The legislation provides protection for the birds, their eggs and nestlings from killing or injury. It also covers the destruction of nests and any intentional disturbance while building or attending to a nest. It is a Red Data Book species (designating a high probability of national extinction). Finally, as a protected species, black redstarts form material considerations in planning decisions. The British Government's *Planning Policy Statement: Biodiversity and Geological Conservation* (PPS 9) (ODPM, 2005) describes policies that need to be

> taken into account by regional planning bodies, in the preparation of regional spatial strategies, by the Mayor of London in relation to the spatial development strategy for London, and by local planning authorities in the preparation of local development documents. They may also be material to decisions on individual planning applications. (ibid.: 1)

In the accompanying circular on *Biodiversity and Geological Conservation* (ODPM and DEFRA, 2005), black redstarts are one of a number of protected species whose presence forms a material consideration in the assessment of a planning proposal.

> The *presence* of a protected species is a material consideration when a local planning authority is considering a development proposal which, if carried out, would be likely to result in harm to the species or its habitat ...
>
> It is essential that the *presence or otherwise* of a protected species, and the extent that they may be affected by the proposed development, is established before the planning permission is granted, otherwise all relevant material considerations may not have been addressed in making the decision ...
>
> However, bearing in mind the delay and cost that may be involved, developers should not be required to undertake surveys for protected species unless there is a *reasonable likelihood of the species being present and affected* by the development. (ibid.: 33, emphasis added)

Given this battery of protective statutes and policies, it becomes important to consider how black redstarts are made present.

While not a shy bird, the black redstart's limited numbers combined with survey problems make recording presence a less than straightforward matter. First, even at first light when survey work tends to be carried out, urban noise levels can make it difficult to discern even powerful and distinctive singers like black redstarts. Second, while background noise is one thing, there is also the bird's tendency to remain quiet during the breeding season, which is the period it is most likely to be in Britain. Third, there is often poor access for people to sites frequented by black redstarts (especially railway lines and derelict properties). Fourth, there are few uninterrupted lines of sight in cities, making visual 'spotting' especially difficult. These last two points make for a fairly demanding landscape for bird surveys. Finding a black redstart may involve climbing walls, trespassing property, running across busy roads and any number of manoeuvres to catch a glimpse of a bird before it moves behind a roof top or through the broken window of a derelict building. Rarely is there a more palpable sense of the different spaces inhabited and mapped by humans and different species as when trying to follow a bird through a city. Possible sightings, lost from view, are followed by sometimes long periods of relative inactivity, waiting for birds to reappear. Caught in the lens, a rooftop bird may be little more than a non-identifiable shape. Once they take to the wing, the flight pattern may help identification, though the narrow field of the lens can make it difficult to track rapid three-dimensional movements. On a perch or in flight, adjustments to a bird's position can provide a different profile or allow early morning light, with its long shadows cast by tall urban structures, to bounce off another part of the bird's body, revealing or concealing distinctive field marks (some of these difficulties are also noted in Law and Lynch, 1990). In sum, the definitive, identificatory, line of sight often involves the movement of both parties.

Even if a bird is positively identified, there are other problems to overcome. Among these are the logical problems of estimating population sizes from sightings (allowing for multiple sightings of the same bird, mistaken recordings, and so on), the difficulty in identifying breeding and foraging sites (for conservation, breeding sites are a priority) and the difficulties involved in recording migrating species (duration of stay can be short for foraging birds, a feature that affects the estimated population size). The mobilities of living organisms underline that presence is as much about intersecting trajectories as it is physical co-presence.

In short, and as for the water voles, recording embodied presence is far from straightforward. Even though black redstarts are relatively conspicuous, they do not present themselves. Fleeting assemblies involving black redstarts and people are necessary. The point to make here is that the survey makes particular relations possible and at the same time is less capable of translating other relations from the field to other spheres of action. In theory, it is an answering machine, able to record presence or absence. In practice, it is a technology that provides glimpses and fleeting presences. Absence from records does not necessarily mean 'not present' and recorded presence may be more or less significant depending on the behaviours of the bird. A dot on a map of the city signifying a positive sighting contains as many absences as it does presence.

Taking these flickering relations together with the socio-legal framework suggests something of a problem for conservation practice. Stated baldly, the rarer a living thing, the more the protection, but the less often it will be implemented, because there are fewer instances where it can take effect. For legal protection is, of course, dependent upon the establishment of embodied presence, on finding something to protect (see the earlier quote from the Government Circular accompanying PPS9). On the face of it, this seems correct. This is a pragmatic spatial mapping that can deal with ecological differences – as long as they are easily marked in terms of presence or absence. However, nonhuman mappings (the spaces of species) and attempts to map nonhumans can take many forms, forms that may elude any legal prescription of either presence or absence. The glimpses produced in surveys and the at best fleeting presence of inhabitants like water vole and black redstarts in the urban landscape point to other space-times. It would seem, then, that the legal topology of presence and absence can, in some circumstances, be unreliable as a means to further conservation.

Enacting likely presence

In practice, and in the UK at least,[1] a procedure has been worked out to cope with these problems of things not being firmly in place. With regard to black redstart conservation, a working category has emerged called 'likely presence', a term that ecologists are now using in their dealings with landowners and would-be developers in the city centre. The process works like this. Once a planning application has been lodged, the city council, following any objections raised by, among others, NGOs like the Wildlife Trusts, may make a 'likely presence' claim. This will be the case in particular where there are known ancestral nest sites. (Knowledge of these sites and the spatial histories of species is both formal and informal, and most often relies upon the longitudinal records of amateur naturalists.) For black redstarts, 'likely presence' has the virtue of being loose enough that it can be applied to most of Birmingham's city centre, and a good part of London's Thames corridor. (It tends not to be invoked in more suburban settings, where black redstarts are largely unrecorded, most probably as a result of significant changes in flora and fauna compared to city centres and competition from the highly territorial robin (*Erithacus rubecula*)). The likely presence claim is enough to require that a survey be carried out. The aim of this would be to establish either the presence or the absence of black redstarts, and if applicable, impose the Wildlife and Countryside Act and any necessary planning conditions (through the Town and Country Planning Act 1990 and Planning Policy Statements) on the developer.

Surveys for black redstarts are labour-intensive (they last anything up to four months, must occur at a certain time of year, and cost a significant amount in terms of the survey itself, and significantly more in terms of lost time, anything up to 12 months, on a building project). Given these costs, and of course the risk for the conservationist that categorical presence will not be demonstrated, parties are more inclined to enter into discussions of how to manage the project in anticipation of possible black redstart presence. The partnership involves an undertaking to orient working schedules and practices, along with building and landscape features, around a likely presence of black redstarts. In addition, developers agree to monitoring of the site during and after completion of the project, and financially commit support, in perpetuity, to the attempt to increase black redstart numbers.

It turns out to be more effective to act as if there was presence, and/or to be seen to be advancing the potential for nesting, or to experiment with different forms of landscaping and building, than to go through the drawn-out surveying and legal procedures that may or may not determine presence or absence. Thus, developers in Birmingham and London are now re-using hard core and other 'waste' as low nutrient substrate for plantings of ruderal species, designing and constructing green roofs and installing relatively intricate nesting boxes (black redstarts, like many other birds, prefer particular built forms which nest boxes need to mimic in some fashion). All of this work is experimental, aiming to generate suitable habitats for black redstarts (and by extension a wide range of invertebrate and plant species).

There are four points I want to make at this juncture. First, there is a sense in which the development of a practical tool called 'likely presence' is an event. It has produced a shift in the socio-technical arrangements, the methodological assemblage (Law, 2004a) of conservation practice and politics. The survey is no longer the paradigmatic device of black redstart conservation. It remains, to be sure, but set at a different velocity relative to the process of forming a political grouping. The survey starts to become less of a technology of adjudication, arbitration or acceleration. Rather, it is now the first, possibly faint, attempt to start a longer experiment. Or, if all else fails, it can be used as a technology of protraction. It won't necessarily provide the answer (presence or absence), but it can succeed in this socio-legal set-up in providing the glimpses necessary for a likely presence claim, or in threatening long delays to development decisions.

Other materialities and technologies have started to become at least as important in this process of disclosing a possible landscape for people and black redstarts (that doesn't require urban dereliction or a war). A notable shift is the growing importance of the biological or ecological record. The ability of such records to act as repositories for long-term observations, often carried out by lay experts with day-to-day knowledge of sites, species and their spaces (Waterton and Ellis, 2004), becomes a crucial element in supporting claims of likely presence. In turn, crushed concrete, sedum mats and other growing media that can be used to produce nutrient-poor habitats on adjacent land and on roof spaces become important. Nails and hammers, wood and projects for school children, all are used to construct the long nesting boxes (which mimic flight into derelict loft spaces). Likewise, new forms of costing projects emerge. Direct and indirect cost savings can be mooted and potentially incorporated into project costings. They include treating what was hitherto considered as waste (land scrape, demolition materials) as landform material, reducing transport and landfill costs and lowering inputs of top soil and aggregates. Other running cost savings include enhanced energy efficiencies and broader benefits including longer residence times for on-site drainage water, thereby making a contribution to the hydrological characteristics of urban developments. Much of this requires complex material adjustments to markets and schedules for building projects. The process is slow, but there are high profile developments

Figure 8.5   Mimicking the derelict – the Barclays Bank Building on the Isle of Dogs, London, is one of a number of projects where a green roof generates a number of environmental benefits. There are ancestral black redstart breeding sites close by.

where some of the 'anti-aesthetic' of this form of urban design reaps cost and symbolic benefits for architects and developers (The Laban Dance Centre in Lewisham, London, with its green roof and brownfield landscaping won the 2003 Stirling Prize for Architecture. In Birmingham, along with Urban Splash's high profile redevelopment of Fort Dunlop to include green roof technology, there are attempts to produce developer partnerships that are informed by coherent strategies for brownfield ecologies in the East Side inner city development – see Figure 8.5).

Box 8.1   Living cities and sustainable cities

There are two architectures of cities emerging here. The first is the more conventional type associated with urban sustainability. It is a city where nonhuman materials figure only as a substrate or resource stream for human livelihoods. It is a form of green city that is reified in architectural blueprints (Jenkins, 2005; Rogers and Gumuchdijan, 1997), where ecology and sustainability figure a fully acculturated nature of 'synthetic materials and energy protocols' (Foster, 2006: 12). It is the sustainability of the UK's Urban Task Force and their urban renaissance (DETR, 1999). It conforms to what anthropologist Tim Ingold would call a building perspective (Ingold 1995; 2000). In

(Continued)

the building perspective, an environment can enter a city story in particular ways. Its components can clearly demonstrate certain properties of transience, stubbornness, elusiveness and even resistance – relative to a specific set of relations which will include how the economy is working and the current form and status of knowledge. Such is the reason that cities are built from different materials and to different specifications. But, even allowing for a certain amount of unevenness in the topography of the nonhuman world, it remains that, within the building perspective, the only significant relations in any account of city life would be understood to be predominantly human-centred. Meanwhile, any nature that is admitted to this kind of account is of a second order. It is socially produced like the rest of the urban sphere. It is the manipulated, recursive nature that has already been pressed through the mangle of representation. While it is fair to say that a good deal of the recent writing on urban natures from a more or less Marxist frame has managed to be wary of some of the worst tendencies of the building perspective (particularly its apolitical technocentrism), there is nevertheless a remaining tendency even here to treat nonhumans as second nature, as produced in and through human labour (Gandy, 2002; Swyngedouw, 2004). While these and other studies are intellectually and politically important, the difference of nature tends to be underplayed. As Braun has helpfully put it, in these studies,

> While there is great deal of talk about the importance of nature to understanding the city, it is often unclear what nonhuman nature adds to these accounts except the presence of a static stock of 'things' that are necessarily mobilized in the urbanization process. (2005: 645)

In place of, or alongside, this sustainable city, there is the living or dwelt city (Hinchliffe and Whatmore, 2006). This is a city of multiple practices, where plans and built forms are different in degree and not in kind (see Chapter 5), where 'people do not import their ideas, plans or mental representations into the world … Only because they dwell therein can they think the thoughts they do' factor (Ingold, 1995: 76). The decentring of human cognition in Ingold's work is helpful as a means to in many more things to cities, and to cede them roles other than as means to pre-set human ends. It figures an openness to others, and allows us to broaden our understanding of what goes into making cities liveable. It is a shift in emphasis that doesn't necessarily displace other architectures (the buildings cited above are only ever partial experiments, and green roofs, for example, soon become standardized material forms), but it can help to prevent architecture from 'turning in self-satisfying circles' (to borrow a phrase from Isabelle Stengers who is talking about the not unrelated field of complexity, see Stengers, 1997: 5). For more on open form of sustainabilities see Chapter 10.

Second, accommodating black redstarts is not founded on a certainty, or a metaphysics of nature or of presence. It is rather precautionary, acting on matters of concern rather than waiting for the facts. While there is little dispute that black redstart numbers are low, and scientists (and I include in this category a whole range of practitioners who contribute to the field) have demonstrated as much through their well-honed techniques in understanding and describing bird life, there is a point at which this knowledge no longer becomes *the* question over which to argue. The trial has changed. The question becomes how can we live together, a question that draws us away from the trial by presence to a trial of converting likely presence into a new cosmo-political arrangement (see Latour, 2004b; Stengers, 1997; 2000).

Third, there are numerous relations being generated in this story, some of which are about drawing things together, others allowing things to stay elusive, absent and uncertain. From identifying black redstarts in the city to the experimentalism of 'new' building practices including brown walls and green roofs, and terrestrial landscaping that avoids the almost ubiquitous bark-mulch-and-berberis schemes favoured by landscape firms, there is a disclosing of possible worlds at the same time as attempts to generate a difference, to form a new collective (which has a wide and semi-permeable boundary that is full of the necessary comings and goings).

The fourth and final point I want to make here is that of course this might be a limited example, an event that may or may not move others, and may therefore be of limited scope (see Stengers, 2000: 67). Accommodation of black redstarts is relatively affordable (and may be cheaper than not accommodating them, given the costs saved on re-using aggregates and in terms of potential hydrological and energy efficiencies that are afforded by green roofs, for example). Economists are involved in costing these and other gains. Even without this demonstration of benefits, there is a sense that the agreement to proceed is itself a *short-cut*, based in this case on economic gains alone and on a kind of green wash for developers. But environmental scepticism is not the only point to be made. Another task for social scientists might be to make it their job to note that we have traced a shift in the set of rules for conservation (as simple as this is, from presence to likely presence). Now it is time to see what else we can do (assembling our onto-political techniques, including economics, geography, and so on, to intervene in practices).

One possibility takes us back to the plight of the water vole on derelict land in city centre Birmingham. Part of the problem here was the difficulty in reading the landscape for signs that are often subtle and transient. But the ensuing contest over presence or absence of water vole on this site (and, thus, the contest over the kinds of development that should take place) might be more productively addressed through a discussion of the potentialities of urban settings. Here we start to move away from that tendency of conservation to render the present eternal (Bowker, 2004). Instead, it's a more dynamic mapping. Rather than seeking the political short-cuts that seem possible through legal structures and a reliance on species presence, there might be a more open discussion of and experimentation in how this site can be remade. And, as I have also argued, it is not simply the likely presence of embodied representatives that is at stake. There can be more to conservation practice than things that speak as specimens.

Ethologies and representation

In bio-philosophical thinking, ethology, or the science of evolution, amounts to a rejection of the notion that living organisms simply perform to an internal script, checked only by external conditions (see for example Ansell Pearson, 1999; Thrift, 2000). Rather the emphasis is on understanding life forms as enfoldings of complex topologies of living and non-living matters (see Whatmore, 2002). In other words, they are assemblages, affecting and affected by their fields of becoming. This tack implies that urban ecologies are urban in more ways than their familiar guise as un-natural or artificial ecologies hindered by a set of obstacles contrived by human technologies and human presence. Rejecting this urban fall from edenic nature, the urban can be regarded as an addition to, rather than always being a sub-traction from, ecological relations (a key reference here is the work of Jennifer Wolch, see, for example, Wolch, 1998).

The question then becomes how to represent these foldings of species spaces. Before I suggest some possibilities, I need to return to representation. As we have seen, representation involves a good deal more than depicting, re-presenting, bringing into presence, recording identifiable identities. It is also an addition, a making, that may be better described as a translation, usefully reminding us that a copy is also a betrayal (Callon, 1986; Serres, 1995a). Which is not to say that representation is destined to be a tragic failure, with its products forever cast as the second class of objects which will never be as bright, fragrant, interesting or complex as the world of the represented (be that a public, a water vole or membership of some kind or other). Rather, representation as translation implies a whole suite of activities which include the loading of the world into words, the construction of a referential chain, the making public of a thing and at the same time the making *of* a public which can *attest* that thing. It is, then, nothing more and nothing less than a *test* of the socio-technical agencement. It is the simultaneous attempt to engage with another, and learn how to engage with another, that is at stake.

For Latour, representation is made up of two powers; the power to take into account and the power to arrange in rank order (2004b: 109). And,

> What is excluded by the power to put in order ... can come back to haunt the power to take into account ... Such is the feedback loop of the expanding collective, a loop that makes it so different from a society endowed with its representations, in the midst of an inert nature made up of essences whose list would be fixed once and for all. (ibid.: 125)

This means that there is a continual process of testing the very constitution that is set in train by the representative process. For Latour, then, the task is not to be forever refining a representation in relation to some pre-existing ground truth, rather, it is to produce representations upon representations with the aim of building a more successful collective.

Birmingham's water voles provide a good example. What the ecologists achieved was both more and less than they imagined. First, despite the use of tracking tunnels, and the careful piecing together of multiple signatures of water vole writing, the evidence for water vole presence was not self-evident. And certainly, their presence remained a contested feature of debates over the future of this site (making likely presence an event that can make a difference). Second, in being alert to more than presence, something else started to matter. Brown rats and water voles were over-writing each other's territories and mappings (in the form of using common latrine and foraging sites), an urban cosmopolitanism that seemed to complicate the predator–prey relation that these species would be expected to perform. It was possible that water voles produced in and through this event were not simply representative of other water voles. They weren't only speaking as water voles. As the field trials proceeded, a sense was generated that here was something that mattered (although why, or to what ends, it mattered was by no means clear) (on this Whitehead-inspired sense of creativity, see Fraser, 2004; Stengers, 2004).

One possibility could be that, in the discussions on how to take this site forward, it was not only water voles in general that should be part of the reckoning, but *these water voles in particular*. This would have an effect on the ways in which plans for the site could proceed. More work would be needed to preserve and enhance the complexity of the habitat, and in particular its interconnectivity for water voles. The water voles might require a multi-layered ecology that allowed them to under- and overwrite potential predators and other hazards, including an urban river that is prone to rapid hydrological responses – a feature that made a boggy area adjacent to the stream, which was earmarked for being drained in order to make way for a traffic island on the development plans, a potentially vital feature of the landscape.

A second possibility would be to treat these water voles as something other than sub-specimen species whose loss could be offset by population gains elsewhere (through some form of planning gain or Section 106 planning agreement). Urban conservationists complain that urban nonhuman populations are seen, by resource strapped conservation organizations at the national level, as surplus to the real business of conservation. The real business being an ethical-rational practice that properly takes place in the countryside (Harrison and Davies, 2002). So while English Nature, the government's statutory nature conservation body for England (and now part of a combined body called Natural England), has a national programme which prioritizes certain species, one of which is the water vole, they were not involved in any dealings with the Birmingham and Black Country populations. But 'these water voles in particular' is a claim of both similarity and difference. The water vole mapping going on at this site and in Birmingham and the Black Country may well be a vital moment for a species that has retreated to the upper reaches of the river catchment. Viewing species as processes, as being reconstituted through overlapping though differentiated populations, whose simultaneous differentiation and compatibility are conditions for its success, starts to complicate conservation strategies. Representation becomes a matter of *more than one* identity, a more complex field of becoming (in this case, of becoming water vole). Representing water vole is, then, more than a matter of recording presence – it involves engaging with potentials, including likely presence as well as differentiated presences.

To end this section, I have started to argue that conservation practice involves various forms of presence. Meanwhile, I have suggested that expert ecologies, tied to legal frameworks and a metaphysics of presence, can reproduce a static ecology which can only make space for the absence or presence of individuals, representing a species, and conceived of as performed homunculi (Ansell Pearson, 1999). The point I hope has been made that ecologies, species and conservation don't always work like that. In the final section I draw out some of the implications of the argument so far for thinking through some elements of a spatial politics of nature conservation.

A careful conservation

As I indicated in the Introduction, nature conservation has tended in theory at least to reproduce a strong distinction between social and natural matters. Nature conservation has therefore tended to vacillate between a political and economic wing which tries to include the excluded in matters of politics (internalizing the externalities, for example) and a scientific wing which tries to save nature from human and political affairs. This chapter's argument has been to suggest that this topology of inside and outside is too coarse. We need to find ways of discussing black redstarts, water voles and many others not as fully fledged members of a body politic or of an economics, nor as perpetual outsiders which must be left as they are, but as things that are not quite present.

One starting point for this discussion would be Latour's reconstituted political ecology, where matters are never settled once and for all (Latour, 2004b). For Latour, any inside/outside relation can only be a temporary arrangement. There is no timeless Nature, rendered as an eternal present or as something sitting mutely outside an evolving society. 'The outside is no longer either strong enough to reduce the social world to silence or weak enough to let itself be reduced to insignificance' (ibid.: 212). The emerging collective, as Latour calls the practical and political realization of a more than human society, is no longer a solid solidarity that blunts any sensitivity to a 'them' (or a world of active human subjects and mute objects). The envelope of the collective is more fragile, the ontologies less certain. Nevertheless things *are* taken into account and put in order. As I have shown, surveys are made, species are ordered, presences are mapped. Latour's point is not that this ordering work should be rejected, but that it should be regarded as one of a series of representations that is an ongoing process. So in his scheme, associations between people and nonhumans like black redstarts and water voles are made or not made, and any such inclusion or exclusion should be regarded not as a matter of timeless certainty, but as a fragile process of moving the collective forward. In future, orderings may be challenged, especially by those that have been left out of the current round of political representations.

While this is useful, and certainly more subtle when you add Latour's previous works to the picture he is developing of a newly ecologized politics (perhaps especially but not only Latour, 2000; 2002; 2004a), there remains a tendency to see entities as either associated or not with the emerging collective (for a similar critique, see Mol, 2002). Yet in practice it may not be a matter of things being (however temporarily) in *or* out. The very rendering of the space of a collective as more fragile than its solidaristic alternatives has effects on the way we imagine the closeness or otherwise of things. There is something else to the collective than inside/outside, more ways of relating than the either/or of association. There may, for example, be accommodations, adaptations, indifferent meetings, and so on (see Chapter 5).

Another possibility for thinking the spatiality differently is to remember that association is, to put it too bluntly, a letting go as much as a bringing together. Rather than a matter for grasp and control, getting nearer to something like a water vole or black redstart can be a realization of distance. Stephen White's use of a Heideggerian sense of distance and proximity in his conception of a political grouping can be indicative here.

> Near and far do not function simply as spatial opposites, but rather as two sides of a playing back and forth in relation to the other. Attentive concern … means that the gesture of nearing, bringing into one's presence, into one's world, must always be complemented by a letting go, an allowance of distance, a letting be in absence. (White, 1991: 67).[2]

This is not quite the same as saying, as Latour tends to, that any constitution of a collective implies outsiders who may be more or less threatening. It is also saying that those inside the collective will be at best partially connected (Strathern, 1991), lending their worlds in ways that may well be far from complete and will therefore continue to surprise.

So, while there is the right of reply for the (possibly friendly) enemies of the collective, the friends of the collective are also far from settled, possibly not present, and undoubtedly on the move. By expanding the spatialities of nature, we can further attempt to disclose a world for people and black redstarts who are likely presences and develop a collective where 'these water voles in particular' can breathe. This amounts to a more careful political ecology, not in the sense of being cautious or even being full of care (in the sense of sheltering others), but in the sense of being open to others, or being curious about others. Contrary to the more humanist elements of thinking about care (including White, 1991), it need not be conceived of as a virtue, a trait that we can somehow cultivate in ourselves, figured as lone bodily projects. In this sense there is a departure here from some of the practical philosophies of Foucault, Rorty and others (Shusterman, 1997). Rather, care is produced with and as others, and is neither selfless nor only about the self. Indeed, it is an ecology that is not oriented to securing an inside (an us) nor oriented to everything outside, but a gathering together that is not too tight and can thereby work to confirm rather than to assimilate others.

Conclusion

Things do not lend themselves to a political ecological project in a straightforward fashion. Associations and collectives are not envelopings, bringing matters into the fold, nor are they matters of binding things tightly as specimens or representations, speaking as this or that, and only that. It's an attentiveness to difference that makes for useful, curious and surprising relations, and for a collective that can stay on a learning curve (Latour, 2004b).

I have argued that natures, or the facts with which conservationists work, are often unfinished matters. They guide action but they do not determine outcomes. Meanwhile, making things present is also a highly contested issue. Taking these two insights together suggests that conservation is not a matter of sheltering, or rendering the present eternal. Sheltering can in this sense be a form of smothering, and, in being insensitive to difference, is a form of incuriosity and therefore cruelty. Similarly, fixing the coordinates of other species amounts to another kind of insensitivity, and is potentially as destructive as not noticing. Certainly, rendering the present eternal, through what I suggested at the outset were metaphysics of nature and presence, does not bode well for black redstarts or water voles. The former might not show up, and the latter, if they do show up, only show up as specimens that are expendable in the world of tradable sites and planning agreements.

Insensitivity to the differences of nonhuman inhabitants can take many forms. It can range from a failure to recognize their mappings and likely presences, to a failure to attend to the differences within and between species. Concern and care involve attention to the details of the lives of others, to understanding that those details matter, even if and especially when why they matter is an open question.

The options for conservation sciences are now, on the one hand, more difficult than they might once have been. On the other hand, they are also more interesting, for the facts of the matter are no longer (imagined to be) on the table prior to practice. The facts of the matter are lures to feeling, to considerations of the adequacy of our experiences to things that matter. The question for conservation cannot simply be about present presence, the gathering up of all that matters, once and for all, and then devising means of rendering them eternally present. Instead of acquiring specimens for a menagerie of the timeless, conservation practice works with natures that are being reconstituted. Their stability is not given before political ecology starts its work. And any stabilities that are produced need to be provisional, working categories, that enable rather than disable further learning and another reconstitution of nature.

How to translate this emerging ethos, this style of knowing which is more akin to an experimentalism than a metaphysics of nature (to use Latour's distinction once again, see Latour, 2004b), to a broader realm of environmental practice? That is a question that animates the final two chapters. I want to move outwards from conservation, and outwards from the case materials I have used so far, to open up more questions for the kind of ethos that can travel with a reconstituted space for nature.

Background reading

Bill Adams's *Against Extinction* (2004) provides a wonderful overview and insight into conservation issues.

Further reading

Further details on the examples and their relation to politics can be found in the following (Hinchliffe et al., 2005; Hinchliffe and Whatmore, 2006). Geoffrey Bowker's work on biodiversity extends many of the arguments here to discuss making data and the currencies of conservation (see Bowker, 2000; 2004). Latour's *Politics of Nature* (2004b) develops some of the issues raised here in relation to the metaphysics of nature versus a metaphysics of experimentalism.

Notes

1. While state and legal apparatus are bound to vary in terms of the degree to which probabilities and possibilities are granted legitimacy, the degree to which this affects the processes that I am engaging with here is an open question. Even in the USA, for example, where legislation tends to be more formal and less tolerant of uncertainties (thanks to David Demeritt for this observation), there has been a recent example of a woodpecker, considered extinct, being made present on audic tape in the form of a 'flicker' that might or might

not suggest its presence. The flicker was enough to temporarily halt construction of a $320 million irrigation project in Arkansas.

2. 'One's presence' etc are more than a little suggestive that Heidegger and White share an interest in the centrality of the human subject in this process. But the argument can be extended to other things and to relations between nonhumans, as Harman (2002) has demonstrated in his non-humanist engagement with Heidegger's work.

Towards a caring environmentalism

The feminist philosopher, Chris Cuomo, has a rather nice way of capturing a recurring problem for environmentalists. 'To what does "environmental ethics" refer ... when our ecological agency is less like a vector and more like smoke?' (Cuomo, 2003: 98). Earlier in the same essay, she offered this description of the current problem:

> More and more, as members of global postindustrial economies, we are in close ethical proximity with people, communities, nonhuman species and ecosystems that are very distant from us, geographically, affectively and epistemically. Our lives are so enmeshed with the lives of distant people, places, plants and animals that it is ridiculous to even pretend that we have an emotional or epistemic connection with our mortal worlds. We are members of economic and environmental communities too large, too diverse to even imagine.
>
> What might it mean to promote the good of a community you cannot even hold in your imagination? (ibid.: 97)

Cuomo is highlighting space and spatiality as a major issue for environmental ethics. We are ethically proximate to people who have not yet lived (think of the effects of nuclear waste), to people on the other side of the planet, to animals, bacteria, carbon molecules ... The list, as Cuomo suggests, is endless, and beyond words. But the ties are real, and being made and remade at a rapid pace and with intensity. In other words, the list of potential concerns or members of an ethical gathering is by no means limited to those living now, those with whom we share capacities (for language, thought or whatever else), or those with whom we share anything much at all. Ethical concern is spatially and temporally complex. How then to proceed, and proceed well, when the constituents are so disparate and unknowable?

This chapter is an attempt to take these questions forward, to make a little progress, given what has been said so far about the multiplicity that marks geographies of nature. I focus on human–animal relations – partly to follow from the refiguring of conservation discussed in the previous chapter, partly because of a motivation to engage animal politics arising from a care for animals, and partly because it forms a useful test for thinking and practising geographies of nature differently.

Human animals – what kind of relation?

Human animals and nonhuman animals are not easily separable, nor are they reducible to one another. They are folded together, materially and semiotically (see Chapters 4 and 5). And through these relations there is alterity, difference in relation (Chapters 5, 6 and 8, for example). But what kinds of relations are to be fostered in order to do animal–human better? And, by extension, can such relational lessons be used to understand and do better by other things (i.e. not just animals like us)?

More than most, perhaps, it has been the work of feminist science studies scholars, and Donna Haraway in particular, that has fuelled the re-imaginings of what might be involved in a politics of/with nonhuman others. Mobilizing monsters, cyborgs and coyotes, to name a few (see, for example, Haraway, 1985; 1991a; 1993; 1997), Haraway has highlighted the numerous human to nonhuman and nonhuman to human transgressions involved in becoming 'us'. Most recently, in her *Companion Species Manifesto* (Haraway, 2003), Haraway extends this work to explore possible alternatives to current interspecies relations. It is an attempt to trace paths to other arts of living. In this sense it marks another step away from the anthropocentrism that inhabits so much social and philosophical thought (on this, see Derrida, 2002; Wolfe, 2003b).

Haraway's manifesto is about dogs and people, as specific bodies in relation. It is specific. But it is also about more than dogs and people. To use Haraway's phrase, it is also about a politics and ethics committed to a flourishing significant otherness (Haraway, 2003: 3). I want to rehearse some of the ways in which Haraway traces specific types of human–dog relations, and through this tease out what such traces can engender for animals' geographies[1] and more broadly for an understanding of geographies of nature. My main aim is to be able to learn, with Haraway's help, as much as possible from these promising human–dog relations in order to start to work out ways of translating this learning to other scenes. These translations of dog–human relations are not straightforward. Nor should they be. Translations are never simple copies, even where they are intra- rather than inter-specific. Haraway is clear on the specifics of species, the difference that this difference makes. Taking this as a central concern, this part of the chapter is an attempt to draw out the possibilities for and limitations of companion species in terms of other topics, other places, where different specificities matter.

A short natural history of kennelization/domestication

For Haraway, we're not, and never have been alone. 'We' live and have lived in and through numerous shared histories. For example, dog and human have been mixed up for at least 100,000 years. We've known this for some time of course, and there are numerous conventional accounts of domestication (or even kennelization) which have celebrated human and canine ingenuity. Human or dog, depending on who you read, use one another as means to their own ends. (Admittedly, in the wider literature, it is mostly 'man' who is given this animal-tool using capability, but occasionally dogs are accorded the same honour, using human beings as means to further their reproductive or other ends – for an example, see Budiansky, 1997 and a review, Anderson, 1997). But what Haraway adds to this mix is something rather different – the idea that domestication was a process like all others that blurs means and ends, that embarks on pathways and traverses the world in ways that could not be wholly known beforehand. In other words, rather than accounts of being with others, of human beings or dogs getting along with the help of their fellow travellers, there is more at stake and more play in the arts of relating. Haraway consistently traces histories of becoming other in relation, or to use a term of Latour's, histories of being-as-another (Latour, 2002: 250). Which is to say that the dog–human pair have not merely learned to use one another, but to explore 'heterogeneous universes that nothing ... could have foreseen' (ibid.: 250). They are implicated in one another's histories. They are entangled in ways that produce others – nothing stays the same.

> Nothing, not even the human, is for itself or by itself, but always *by other things* and *for other things*. This is the very meaning of the exploration of being-as-another, as alteration, alterity, alienation. Morality is concerned with the quality of this exploration. (ibid.: 256, original emphasis)

Indeed, the power of Haraway's argument is in the refrain that humans and dogs are conjoined in a dance of being-as-another. It is a dance that moves to numerous rhythms and temporalities (from millennial shifts in physiognomy to second-by-second responses and pre-intentional adjustments to one another). The dances are of course far from over and, as we will see, there are continuing possibilities for reconfiguring ourselves-as-others.

For Haraway, then, history is driven by relations, by beings and matters related as they are by their excretions, by texts, by events that are beyond their immediate ken. She traces a knot, in other words, of partial connections that mould, model and are moulded and remodelled as subjects in relation entangle and disentangle themselves. The wonderful term used for all of this knotted cat's cradle of activity is *metaplasm* – the remouldings that inheritance and other relations produce in the stuff of acting out relations. Metaplasm is a technical term for the changing form of a word (through addition, subtraction, transposing letters, and so on). It also has a pleasingly biological tone, reminding us of the additions, subtractions and transpositions that can occur in the codes of the living (from their DNA to a whole variety of relations) when conjoined or conformed (see Chapter 6) bodies share their daily bread (or become com-panionable). Metaplasm is a term that Haraway applies suggestively to words and biologies, to flesh and signifiers, to stories and worlds (Haraway, 2003: 20). It is an elegant and suggestive term for starting to explore being-as-another, sensitizing us to the multiple transfers and translations that make inhabitation possible.

Driving all these metaplasms is a world of unimaginable complexity. The conjoined histories of human and canine are, as I have mentioned, practised and performed through a multiplicity of times and time-spaces. As Haraway puts it: 'The world is a knot in motion … A bestiary of agencies, kinds of relatings, and scores of time [that] trump the imaginings of even the most baroque cosmologists' (ibid.: 6). On the one hand (and in the hands of so many authors), it's a bewildering, bewildered world. So many baroque or other complexities can have the habit of instilling political inertia (for exceptions, see Law and Mol, 2002). But on the other hand, and in other hands, this is an invitation to engage the world differently, to work for a reconstitution of naturecultures. So how to engage? What kinds of relation are needed. Do they need to be codified, enshrined in law and/or treated as rights? Or are we more interested here in affective registers? Or both? Again Haraway offers numerous resources for working this through.

From rights to relations, to specific kinds of relations

The first thing to note is that Haraway tends to side with the critics of abstract formulations of animal rights. She sides with those feminist scholars and practitioners who find the utilitarian approach of Singer (1984) and the rights approach of Regan (1984) (the dominant movers in the animal liberation literature) to be problematic in terms of their exclusions, essentialisms and their normative rather than affective attempts at extending liberty (see Box 9.1).

Box 9.1 Ethical animals

There are three common critiques of animal rights approaches, most of which have been developed first and with the greatest clarity in feminist and ecofeminist literatures (see, for example, Slicer, 1991):

1 *Normative vs affective approaches*: It is sometimes useful to imagine that there are two aspects to ethics. The first can be characterized as a moral code – a guidebook that says in such and such a situation this is the right thing to do. The second aspect is just as important but can often be overlooked. It is the need to develop a form of sensibility or style which is

(Continued)

(Continued)

amenable to ethical questions, which generates the impetus to enact a code. So there are two questions. First, what kinds of codes do we need to live by? And second, how can a sensibility to ethics be cultivated? The first is amenable to reason, to working from ends to the means to meet those ends. It is the calculative realm of action. The second requires some catalyst unsusceptible to full understanding – unthought or bodily regimes, high art, and so on. Stress on the first kind of question produces normative approaches to ethics, while the second is a more affective approach. It is generally argued that conventional accounts of animal rights and ethics have focused on codes and have underplayed the affective dimensions involved in human–animal relations. They have therefore provided few resources for understanding why it is that people feel moved to form ethical relations with others in the first place (Bennett, 2001).

2 *Exclusions*: Extending liberty, rights, etc. to animals tends to do little to the overwhelming logic that human beings stand as the supernatural beings who can recognize some elements of sub-humanity in those species that have until recently been viewed as nonhuman. "Higher" mammals are normally the first to be included in this zone of rights. But any such inclusivity, based as it is on a checklist of possessions, capabilities and/or capacities (be that an ability to use language, to suffer pain, to encounter fear, to have interests ...) always risks problematic exclusions. Do we exclude some human beings from the enfranchised simply because they do not use language, do not seem to feel pain or articulate interests, and so on? And what of all those organisms which don't quite make the grade? Such questions have led to some seriously worrying debates for animal rights protagonists, but they also fail to deal very well at all with the realization that nothing and no-one exists on its own, for itself – indeed, a relational view of any organism or other entity starts to unsettle the notion that a boundary can easily be drawn around rights bearing individuals of whatever kind. To do so is to exclude all the matters and other organisms that make that individual a possibility (Whatmore, 1997).

3 *Essentialisms*: In focusing on certain essential characteristics that are required in order to be considered for membership of a moral community (in Regan and Singer's case, it is interests that are definitive), there is a tendency to first of all play down, or render inessential, all the other matters and aspects that go into being and becoming. Members of the community become, essentially, organisms that have pre-set characteristics, and are, in this case, not much better off than Cartesian automata who already have their preferences and interests mapped out for them. In other words, there is no space for creativity, for life, for natures. Second of all, the resultant

(Continued)

community is also defined by its sameness and not by the complex of differ-ences that are important in making any grouping viable. There is then an indifference to difference built into these abstract animal rights literatures.

These often-made criticisms, which all tend to focus on the impoverishment of worlds, once such abstract, timeless and space-less principles are invoked as a means to secure animal welfare and rights, are powerful. But it should be added that Singer and Regan's works, their academic and activist endeavours to put animal welfare issues on political and intellectual maps, have done a great deal for animals in a variety of settings (Slicer, 1991). The animal rights movement, whatever it can now become, owes a good deal to their endeavours. This debt has most recently been picked up and developed by Derrida (2002; 2003) (see later in this chapter).

In articulating an alternative to abstract rights, Haraway follows and extends the work of the late horse and dog trainer and anti-rights campaigner Vicki Hearne (1991). Haraway summarizes:

> The question turns out not to be what are animal rights, as if they existed pre-formed to be uncovered, but how may a human enter into a rights relationship with an animal? Such rights, rooted in reciprocal possession turn out to be hard to dissolve; and the demands they make are life changing for all the partners. (2003: 53)

In this view, rights do not pre-exist the relationships of human and animal. Dogs and humans claim rights against one another in this face-to-face, body-to-body dance. It is, in other words, through the interplay, or playing off from one another, the moves, actions, reactions and responses, through what Cussins (1996) called an ontological choreography, that the arts of domes-tic/kennel living are explored. For Hearne and to an extent for Haraway, the dog/dog trainer relationship becomes the paradigmatic case for this learning to live the good life:

> Much companion animal talent can only come to fruition in the relational work of training. Following Aristotle, Hearne argues that this happiness is funda-mentally about an ethics committed to 'getting it right', to the satisfaction of achievement. (Haraway, 2003: 52, original emphasis)

Rights, then, are in the relation, in the intersubjective encounters and in the sharing of life changing experiences. It is the relation that matters and makes matters matter. For Haraway, 'the relation is the smallest unit of analysis, and the relation is about significant otherness at every scale' (2003: 24).

This observation is of course consistent with much of the argument of *Geographies of Nature*. Dogs, people, rights, natures – all emerge in practice and through relations. What is perhaps most useful here is that Haraway is concerned not only to differentiate relations but also pass some judgement as to the kinds of relations that will produce better ways of living. More explicitly, she is concerned to distance herself from several relational types that are imported into conceptualizations of human–dog practices. One is the master–slave relation. Another is the parent–child. Yet another is a dependent relation. As we will see, none of these describe the companionable relation that Haraway is after.

As the language of reciprocation and mutuality suggests, companion species is not meant to be read as synonymous to a master–slave relation. There are at least three reasons for this. First, there is play and uncertainty in the relationship. This is not about control. Indeed, the idea of mastery is surrendered for it is only through being-as-another that the quality of the relationship improves and affords successful training. To put this even more baldly, the idea of mastery has had its day in what is now a so thoroughly entangled world that no-one or no-thing can be said to be in control (see also Latour, 2002; 2004b). Second, this is not a question of power moving in one direction, or of a simplistic 'power over' calculus. 'Love, commitment and yearning for skill with another are not zero-sum games' (Haraway, 2003: 61). Human and dog gain and add to the world. The third reason for eschewing the familiar master–slave relation stems from a practical observation concerning the life-forms that have emerged out of these interspecies histories. Master and slave assumes possible freedom for the enslaved party – but our entanglements with dogs and other companion species would turn so-called freedom into near certain death. As Ian Hacking notes, 'their lives are with us, most would die if not in a year, in a generation or two, without us. Packs of domestic dogs gone wild make a lot of noise but do not do well' (2000 : 24). So as for other nature-cultures, there's no practical possibility of a return, just as there is no place to return to. 'We' are, and have always been, dog people and people dogs. The past and the future are dogged. The master–slave dialectic fails to live up to these open-ended (which does not mean pain-free or death-free) and multiple (you could say un-original or non-foundational, see Chapter 5) relations. To be sure, the spoils of the relation are far from even; there is plenty of bio-power in bio-sociality. Both Hearne and Haraway emphasize the requirement for hierarchies in dog worlds and in human–dog relations. 'Intersubjectivity does not mean "equality" a literally deadly game in dogland; but it does mean paying attention to the conjoined dance of face-to-face significant otherness' (Haraway, 2003: 41).

In a similar vein and following on from this, Haraway steers us clear of the equation of child-love with puppy-love. Again, while it is relations that matter, Haraway is after specific kinds of relation. Puppy-love does little for children or dogs, and can be deadly. The nurturing of dogs and children is different, and we shouldn't assume that ethical extension follows from the kinds of care relation that are sometimes represented as parenting (not least because of the tendency to essentialize particular versions of mothering and intimacy). Meanwhile, there is clearly more to dog–human relations than servicing needs. Relationships of material dependence, while crucial at certain moments, tend towards treating animals as child-like, and needy, and lead more or less seamlessly to matters of provision and/or to sentimentality. Dependence is not, then, the most promising form of relation from which we can remould our world.

In contrast, it is the relationship of mutual affect that promises most for Haraway. The conjoined dance of face-to-face otherness is where we need to start. The training relation provides a rich case. The giving and taking, the co-performances, the routines and the surprises that develop in this kind of relation where both parties become different to themselves, spark the imagination as to how to live well with and as others. Such a dance is of course far from innocent or disconnected from histories and geographies of species and associations of various kinds. So, to be clear, to train and to train well, is not simply a matter of only paying attention to present company. Haraway is clear that we need to reach beyond any particular moment to make all manner of partial connections and naturecultures present.

> When I stroke ... Willem [a Great Pyrenee], I also touch relocated gray wolves, upscale Slovakian bears, and international restoration ecology, as well as dog shows and multinational pastoral economies. Along with the whole dog, we need the whole legacy, which is, after all, what makes the whole companion species possible. Not so oddly, all those wholes are non-Euclidean knots of partial connections. Inhabiting that legacy without the pose of innocence, we might hope for the creative grace of play. (Haraway, 2003: 98)

Only by so doing, only by developing this mode of attention to a complex present, a multiplicity (Mol, 2002), can we be said to be generating naturecultures that can flourish. Through understanding the historical and geographical specificities of relations, of species becomings, the possibilities of the relation open up. Note these histories and geographies are neither determining nor are they endlessly open to re-interpretation. They matter to the degree that dogs are not the same as human children, and retrievers are not guardian dogs.

In the next section I will explore the possibility for translating this particular formatting of promising relations to a more general theme of flourishing.

Translating the manifesto

Does this dance work in other settings? Is it a useful means to work out a way of doing ontological politics? Let me raise three questions – they concern rights, power and space.

First, as I have mentioned, rather than pursue abstract and universal animal rights, Hearne and Haraway attend to the promise of an ethical relation dependent upon exchange and reciprocity. And yet, there are doubts as to the extent to which expertise in training relations can generate an animal politics or ethics: '[While] such work seems to teach us about the complexity of animal phenomenology and subjectivity ... the ethical implications opened by that new knowledge ... appear, strangely enough, to be severely attenuated at best' (Wolfe, 2003b: xvi). The problem for Wolfe is that this cementing of an ethical relation between co-present bodies-in-relation does little to the cartography of ethical boundaries:

> It is not at all clear ... that we have any ethical duty whatsoever to those animals with whom we have not articulated a shared form of life through training or other means. Hearne's contractrarian notion of rights only reinforces the asymmetrical privilege of the ethnocentric 'we', whereas the whole point of rights would seem that it affords protection of the other exactly in *recognition* of the dangers of an ethnocentric self-privileging that seems to have forgotten the fragility and 'sketchiness' of its *own* concepts, its *own* forms of life, in the confidence with which it restricts the sphere of ethical consideration. (Wolfe, 2003a: 8–9, original emphasis)

What Wolfe is gesturing towards is some kind of acknowledgment of the utility of more abstract rights. Perhaps more usefully, we may need to conceive of rights as neither universal and abstract nor derived on the hoof (so to speak). I will return to this in a moment. For now it is also important to register the doubts over, first, the degree to which an orientation to companionable relations can decentre, or avoid the re-centering of, human beings on the ethical centre stage. And second, the extent to which a failure to decentre human being circumvents an attempt to live as flourishing others, to make, in the terms used here, spaces for nature.

The second issue that troubles a translation of companion relations to other matters concerns power. Haraway's text is full of positive sum experiences of human–animal relations. Abilities and capabilities are realized through action. All of which is wonderfully empowering for dog-people, and for people-dogs, for horse-people (Patton, 2003) and for lots of medium to large people-fauna. But there are of course lots of places and others where this kind of power game is less applicable. Here is one possibility.

One of the most striking elements of the novelist J.M. Coetzee's writing is his ability to deal with non-power (see especially Coetzee, 1998; 1999; 2000). In his often revisited scenes of animal death on farms, the killing of stray and unwanted animals and the keeping and killing of feedstock, a sense is conveyed not of human power over animals, or human power with animals, but a non-power (for a similar sense in the foot and mouth crisis, see Chapter 7 and Law, 2006b). Commenting on one scene where a little boy

(Coetzee himself) describes the farm yard killing and disembowelling of a sheep, 'whose insides are just like mine', Ian Hacking notes:

> When you've read a lot about animal rights this comes as refreshingly non-intellectual. No talk of the interests of the lambs being infringed, or the rights of the sheep being denied. Were it not for some tedious puns, one would say that these scenes are wrenching at a gut level; they stick in the craw and bypass the intellect. It is the boy's body, and his feelings of identity with other bodies, that are at work. (2000: 20)

There's an ethical response here, one that is uncertain to be sure, but it figures an emerging ethics nonetheless. But, unlike in the *Companion Species Manifesto*, and unlike some of the animal rights literature with its focus on interests, non-human ethics is in this version of affairs not so much about ability but not-being able. This is an argument, among many others, taken up by Derrida in his engagement with animals (Derrida, 2002; 2003). Elaborating on Jeremy Bentham's famous question regarding animals 'Can they suffer?', Derrida affirms that any such question supposes that 'Being able to suffer is no longer a power, it is a possibility without power, a possibility of the impossible' (Derrida, 2002: 396). This non-power, which is at the heart of power, haunts any attempt to foreground ability or human–animal conjoined capability. Non-power is where

> mortality resides ... as the most radical means of thinking the finitude that we share with animals, the mortality that belongs to the very finitude of life, to the experience of compassion, to the possibility of sharing the possibility of this non-power, the possibility of this impossibility, the anguish of this vulnerability and the vulnerability of this anguish. (Derrida, 2002: 396; Wolfe, 2003a)

What Derrida is characteristically alerting his readers to is that any focus upon the conjoined capabilities of people and animals, or the shared facilities of companion species, must also in the very act focus on the possibility that such powers will cease. For non-power is at the heart of power. And it is this shared vulnerability that Derrida argues is at the heart of the discourse of animals' rights. So, in addition to the question 'Can they suffer?', Derrida would have us consider the relations that are formed around animal death. The practical issue, which is largely left out of Derrida's account, is that as the techniques of killing, in laboratories and slaughter houses, become ever more sophisticated, Bentham's question offers less guidance. With the use of anaesthetics to produce 'humane endpoints' for lab animals, for example, animal ethics may well need to develop strands that both Haraway and Derrida develop, with their combined focus on capability and its absence, incapability. Indeed, while animal rights remains a contested and problematic discourse, it is as dependent on shared inability as it is on ability, figuring 'a new experience of this compassion' (Derrida, 2002: 395).

In response to the irresistible but unacknowledged unleashing and organized
disavowal of this [animal] torture, voices are raised – minority, weak, marginal
voices, little assured of their discourse, of their right to discourse and of the
enactment of their discourse within the law, as a declaration of rights – in order
to protest, in order to appeal … to what is still presented in such a problematic
way as animal rights, in order to awaken us to our responsibilities and our
obligations with respect to the living in general, and precisely to this fundamental compassion that, were we to take it seriously, would have to change
even the very basis … of the philosophical problematic of the animal. (ibid.:
395, original emphasis)

Haraway's work is of course compassionate too, and is perhaps more practically aware of both the promise as well as the hazards of acting as historically constituted compassionate beings in relation. Her short account of
the Save a Sato (Sato is Puerto Rican for street dog) adoption schemes in
North America is attentive to the requirement to take the 'emotional bonds
and material complexity' (Haraway, 2003: 90) of such a scheme seriously.
While efforts to reduce the suffering of street dogs are indicative of a caring attitude to dogs, such a scheme is not easily read as a salvation story,
indicative of a new ethical regard for companion species. Haraway is con-
scious, for example, of the 'racially-tinged, sexually-infused, class-saturated
and colonial tones and structures' (ibid .: 89) that inhabit this scheme. The
transformations of canine bodies (neutered, wormed and doctored in
various ways as they reach the domestic spaces of the 'Northern' dog lover)
recall Tuan's (1984) understanding of the fine line between domination and
affection. Indeed, Haraway is as ambivalent as Tuan with regard to pet
keeping in general when she admits that she is inspired and disturbed in
equal measure by the Save a Sato operation.

The broader point for our purposes here is that any focus on rights as
performed in relation, achieved through a training or other relation, is also
necessarily a recognition that such a relationship can cease, that abilities and
conjoined capabilities can also become inabilities. Shared fragility and vulnerability are as important to interspecies relations as capability. Compassion,
a term that can be used to mark this understanding of fragility, whether
it is shared or not, is, however, far from pure. Compassion, as Haraway
demonstrates, is not innocent. What her more empirically informed analysis
points toward, perhaps, is that human–animal relations are heterogeneous,
made not simply of human and dog bodies, but injections, tablets, international aid, fleas, viruses – the list can of course go on. So any emerging
animal ethic is not easily abstracted from this list of associations. Without
offering any easy answers, Haraway points us to a continual and careful
exploration of possibilities – a vigilant process of relating and finding
better relations, without ever thinking that the problem of how to relate
and relate well has been solved.

Our third question concerning translation involves spatiality. One of the key questions for developing an ethics and politics of flourishing significant otherness is to work out the spatialities of being-as-another. Haraway provides, as I have hopefully made clear, a relational sense of history and geography which invites new forms of intervention in the writing of nature-cultures. The manifesto seeks an ethic, or mode of attention, which is alert to the partial connections, the knot of motions, the multiplicity and the non-Euclidean spaces that make species. The language and topologies will be familiar to those who have followed the anthropological, STS and Science Studies literature (see Chapters 4 and 5, and Mol and Law, 1994; Murdoch, 2005; Strathern, 1991). The practical effects are well demonstrated. Only by understanding the crumpled trajectories of species in relation can we under-stand how best to engage with dogs. Guardian dogs are not herders, nor are they retrievers. No amount of training will change their historical ecology in a single lifetime. Respect for a breed's histories, and geographies, its relational being-as-another, is essential if we are to 'get it right'. Respect requires more than a linear genealogy, tracing the purity of breed, but a sensitivity to all manner of relations.

Haraway uses this mode of attention to understand and practise relations with others that allow for a mutual flourishing. And as I have suggested, this works extremely well for re-thinking companion relations with dogs. But what of other relations with other others? There's little doubt that getting a feeling for another organism (to paraphrase biologist Barbara Mclintock) (Keller, 1992), acquainting ourselves with the particulars of the world we affect (Cuomo, 2003: 103), can be politically and ethically productive. In previous chapters I have been arguing as much. Closeness to prions was a pre-condition for their difference to emerge. Closeness to water voles, getting a feeling for their activities in Birmingham, was part and parcel of an enchant-ing encounter (Bennett, 2001). Getting close is hard work, and is of course as different for prions as it is for water voles. For Cuomo, and for Haraway, it enables 'us' to understand distant relations and 'create alternatives to our present habits' (Cuomo, 2003: 103). Getting closer is also of course about recognizing difference and distance, of allowing otherness to flourish (see previous chapter). Contrary to the essentialisms of mainstream animal rights work, it is to understand the differences that are productive of and produced in any act of closing.

This all seems eminently worthwhile, but there are at least two further spatial elements to be considered. First of all, it is important to be aware that 'getting close' may not be the same thing as 'being in the presence of'. As I have noted, the paradigm case for Haraway is the close, face-to-face, body-to-body relationship of dog and dog trainer. This is perhaps justifiable for companion species. But for others, relations are rarely about co-presence. Many conserva-tion schemes, for example, have to deal with at best a flickering presence of other species, or even presence of absence (see previous chapter). Political and legal frameworks and modes of practice are increasingly based not on the presence of species but on a likely or theoretical presence. Getting close to these organisms is an ontological choreography, but the dance is often conducted once or twice removed. None of this is contrary to Haraway's work. Indeed, she is clear that even what seem to be immediate emotional affects of being-as-another are of course produced through a whole set of relations. But it is worth emphasizing Bruno Latour's oft-repeated point that learning to be affected requires more rather than less mediations (Latour, 2004b). As Haraway would of course concur, this is not about stripping down to our skin and fur, but about finding ways of closing *and* distancing. And the broader point is that an ethics and politics committed to significant otherness may require ontological choreographs where the dancing partners are rarely if at all co-present.

The second point on spatiality is perhaps the most obvious. Lots of brushes with others are presented to us in ways that make light of their existence. For example, the unwrapping of pre-packaged meat products is, for many, the most frequent forms of human-animal relation. How can we translate the ethic of getting it right to conditions where co-presence is of a different order, is mediated differently, or is not achieved at all? I am thinking in particular of a consideration of how distances are made and unmade between species (see Murdoch, 2003, for suggestions). How the close at hand can remain distant, and how we can be open to the far off. This is a topological landscape that Haraway is of course alert to, but which we might mobilize to think through our relations and our being-as-another. Even further removed are those who are not simply difficult to imagine, but the unimaginable connections that produce all manner of effects in all manner of worlds (returning us to Cuomo's concern expressed at the start of the chapter).

The corollary to this is how such distances and proximities affect ethical relations and what work is to be done in developing a spatially multiple ethics. Again, it may well be that the discourse of rights, which Haraway reconfigures in the light of intimate relations, needs different reconfigurations in situations where intimacy is harder to achieve. As Wolfe and Derrida suggest, there may be more in the rights literature that destabilizes the comfort zone of co-present, mutually acknowledged relations. We may need to be alert to the possibility that while 'getting closer' to other species is sometimes possible and can produce desirable, ethical, effects, there are numerous ways in which distances are continuously being created.

Conclusion

There are no easy answers. But it is useful to draw out two points that we can take forward into the final chapter. First, living with others is not simply a call to extend rights to others, who then become part of the community, sharing similar basic features to the already constituted community of human beings (see Box 9.1 on ethical animals). To be sure, codes and rules may well be necessary, a point I will return to in the final chapter, but such formalism should not serve to conceal the differences that are the condition for and the product of relations with others. Differences do not become less important as we learn to live together. Differences grow as we learn to get close. This is the paradox of geographies of nature. Things become more real, more objectionable, increasingly differentiated, as we learn more about them, as we increase the number and heterogeneity of matters that enable us to get close. That is, their differences grow. Developing closeness to companion species, urban wilds, rivers, is also to recognize differences between and within any given setting. Living with others is partly a matter of learning to understand our codependences, our coevolution, but also to respect their differences from and indifference to 'us'.

Second, living as others requires a form of vigilance to the differences that emerge in action. That is, we are not simply different to others, 'we' are also different from ourselves. Another way of saying this is that matters will never be settled, that actions will be overtaken by events. The result of this ontological drift, or swerve, is that codes and rules are likely to become obsolete or at least start to unravel as they are overtaken by events (Latour, 1999). So, for example, the codes that have been generated around the moral question 'can they suffer?' start to become less and less useful once humane end points become adopted across animal research laboratories and abattoirs. We are in process, and thereby different from ourselves. The result is that the politics of animal care, or even of the 'environmental communities too large and diverse to even imagine' (Cuomo, 2003: 97), will remain, in part at least, open.

But codes and calculations are not matters that we can discard at will. As I have intimated, no matter how problematic, the abstractions of animal rights have travelled and made enormous contributions to the task of making spaces for nature. So how to work out a politics of codes and affect, of formalisms and delight, of similarity and difference, of proximity and distance? These questions not only refer to the making of animal politics, they speak more generally to environmentalism, and to the seemingly thankless task of co-ordinating environmental actions in order that they can loosely cohere into a programme, or policy. The last chapter expands on this issue of what it takes to form a loosely co-ordinated environmental policy.

Background reading

Donna Haraway's (2003) *Companion Species Manifesto* forms the main inspiration for this chapter. Like most of Haraway's work, it's a rich resource that is worth reading both as background and then again as further reading. For more background on the contribution of feminist theory to animal and environmental ethics, see Deborah Slicer (1991).

Further reading

Cary Wolfe's work is exemplary of the new animals literatures, and the edited collection *Zoontologies* (Wolfe, 2003c) includes a wonderful collection of chapters that take these arguments into more deconstructive terrain. For an evocative rendering of animal rights debates which highlights formal versus affective dimensions, see Coetzee's *The Lives of Animals* (1999). Nick Bingham's (2006) paper, 'Bees, butterflies and bacteria', provides a very useful review of some of implications of moving from a 'community of brothers' to 'a community of others'.

Note

1   It is more conventional to refer to animal geography, or animal geographies. On this body of work see the ground-breaking volumes of Wolch and Emel (1998), and Philo and Wilbert (2000). As the diversities of these studies suggest, it is perhaps time to pluralize both terms to reflect not only the multiple geographies that animals enact, but also the multiplicity of animals. On the necessity to avoid 'the animal', see Wolfe (2003b) and particularly Derrida (2002).

10 Environmental policies and sustainabilities

Previous chapters have tended to foreground openness. I have been arguing that we need to keep open or even open up spaces for nature. There's an otherness, there are differences in the making, that we ignore at our, and their, peril. From biosecurity to companion species, there is a dance or ontological choreography that discloses worlds that are more than human. Such a call for openness is of course made partly on account of the long history of closed practice and thinking when it comes to nature. As a matter to tame, control, eradicate, bind, secure – nature has largely been left for dead. My argument in this book has been different. Spaces for nature are not pre-existing matters to be circled and shored up, but matters to come, things to be made. They are the fragile end points of actions, the mammals and birds that emerge from fieldwork, the micro-organisms that gain reality in laboratories and on farms, the companion species that flourish in relation. And, even as end points, they may be more than one thing. As matters being done, they will be enacted or worked in various ways, often with the result that they don't fully cohere as a singular object (Mol, 2002). Their complex, multiple, spaces emerge through complicated, heterogeneous practices, wherein matters variously withdraw, associate, make a difference and demonstrate indifference.

A recurring question in the book has been, how to assemble all these practices, places, people and things in ways that are progressive? In other words, how can things be coordinated in ways that don't do untold damage to the very things that are being assembled? In the terms that I want to explore in this chapter, how can a policy or programme of action be formulated that is coherent enough to guide matters forward but loose enough to allow for difference? It is a question that animates the book. How to form a collective without crushing those collected, without restricting what I have called spaces for natures? Other chapters in the book have raised the issue through case studies of agricultural practices, disease control, water vole and black redstart conservation and living with and as other animals. Here I want to go a step further by looking at an example of how environmental policy is being made in the UK. My argument is that policies are also practised in lots of different ways, in different places and with different people and things, and to pursue better policies we need to consider how these various practices are assembled. The term I use for this assembling is 'ecologies of action'. Only by working with the multiplicity of things, paying due attention to the ecologies of action that make things happen, can a policy be said to be sustainable.

Making things concrete

An axiom of environmental policy thinking in recent years has been to 'act locally'. People should be doing it for themselves. It's no good imposing policies from above and hoping that everyone does as they are told. Rather, policy works by enrolling actors in the process. The usual metaphors are empowerment, participation, local action, and so on. In this line of thinking, civil society is the locale for environmental action, not the state. The rationale is rolled out in a variety of settings, from economic development projects, international wildlife and resource conservation programmes, to stakeholder meetings on the consequences of introducing or trialling genetically modified crops to discussions of locations for nuclear waste. In this vein, and in relation to the UK's urban environmental policy, the body set up to advise government on the improvement of urban green spaces, called the Urban Green Spaces Task Force (itself a response to the Urban Task Force which had very little to say on the importance of green spaces in cities, see DETR, 1999 and Chapter 8), noted the following:

> Change is often being led by voluntary and community groups, … Once voluntary and community organisations are involved, they can achieve an enormous amount in articulating local needs, creating or renovating community green spaces, and getting involved in managing traditional parks. (DLTR, 2002: 71).

At face value, this suggests that environments can be organized locally, through civic actions and in accordance with local wishes. And yet, in practice, things are not so straightforward. Let me problematize this view in three ways, and in doing so introduce the case study, 'Concrete to Coriander', an urban cultivation project wherein women living in inner city Birmingham came together to claim parts of the city for cultivation, in the process, greening the city, reducing their social isolation, improving their health and increasing their access to open spaces.

1 A COMMUNITY HAS TO BE MADE, OFTEN ALMOST FROM SCRATCH.
Society doesn't exist ready and waiting to be asked to contribute to saving the world. Rather, groupings need to be fashioned.

Small Heath and Saltley (Figure 10.1), in East Birmingham, are inner city areas that are in the bottom 2.5 per cent of enumeration districts nationally in terms of multiple indicators of deprivation. The population is defined as 80 per cent minority ethnic, suffers disproportionately in terms of health and social exclusion and is recognized as poor in terms of environmental quality and access to green open space. Women in particular tend to be isolated and suffer some of the worst health problems (for more details on the area and the case study, see Hinchliffe et al., 2006). How, then, to bring people and things together in

(Continued)

such a setting? The project started in the offices of CSV (Community Service Volunteers) Environment, an organization that had located in Saltley in the 1990s, taking advantage of Single Regeneration Budget (SRB) funds. City gardening seemed to offer a chance to make environmental improvements in the form of opening up disused or underused green spaces and at the same time contributing towards health and social inclusion initiatives in this part of the city. Given that it was particularly women who suffered social exclusion and had disproportionately high levels of cardio-vascular problems, and given the understanding that people at CSV Environment had of the gender relations that existed within the largely Muslim population of Saltley/Small Heath, the aims would possibly be best realized through a women's gardening club. But the club would not grow by itself. The women were not ready and waiting, there were precious few gardening tools, land needed to be cleared and agreements reached on gaining access to potential plots. In other words, relations of many different kinds needed nurturing. One way to do this was to apply for funds to employ a project officer whose principal task would be to develop the women's group, and gain access to land, organise tools and any labour that was needed to make city cultivation a reality.

So things don't just happen, by or for themselves. There's the work of gathering to consider. This brings me to the second area that troubles the Task Force approach.

2 Activity is dependent upon linking up with other organizations

Financial resources are often needed in order to make things happen. Funds need to be applied for and managed. Such funds are often not generated locally, particularly in areas of socio-economic deprivation, areas where environmental problems are of course often most acute. And money is often 'given' with fairly strict conditions applied. A contract is entered into, and the returns on investment are specified so there can be no doubt that the money is spent and spent well.

In order to secure the required funds, the Concrete to Coriander project needed to be ordered in such a way that it met the criteria of those who would supply the funds. Here's Neil, CSV Environment's Director (NB: Neil is a pseudonym):

It's a bit of a juggling act ... at the end of the day I'm obviously interested in the quality of the work that we actually do and how responsive that is to community need. But obviously I need to recognize that in terms of getting funding I'm going to have to actually

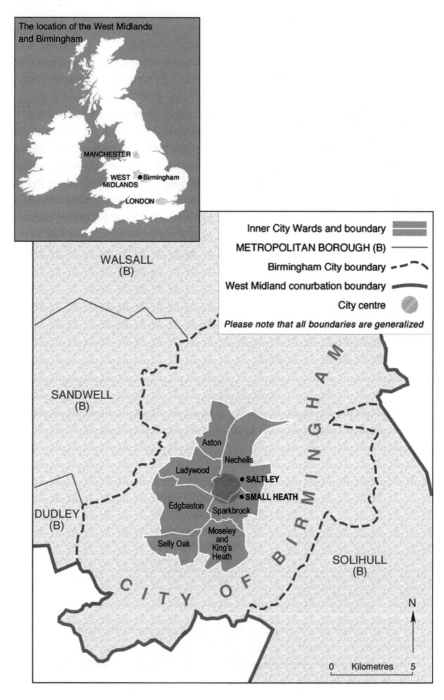

Figure 10.1   Location map of Birmingham and inner city wards

satisfy the needs of the funder. So it's a bit of a balancing act really. To some extent it's actually playing the game or at least the funder's game but trying to come up with quantitative outputs that are not actually going to cripple the project. Because I think that's the great danger. There's always a temptation that the more figures you can write in, the more chances at the end of the day that that will provide you with the money.

So projects start to change as they are translated into the terms of the funding agencies – or at least certain things have to be done and be shown to be done in order to secure funds, and then to release those funds after each period of audit. One way of talking about this is to say that the activities are being framed (Callon, 1998a; 1998b). That is, some things are being made more important than others, some things are being made to count, while others are placed into the background (see also Chapter 6). Which is also to say that certain kinds of space, time and materialities are being written into the project – or at least those parts of the project that circulate on forms, that move from the field to CSV Environment's offices, and then to the regional and national headquarters of funding agencies.

In the Concrete to Coriander case, the project could be judged in terms of how many gardens and gardeners it produced. The application forms tell a story of how the women would develop increased 'capacity' to affect change, they would gain certified knowledge and language skills. All of this would take place in more or less measurable ways over a set period of time (three years).

All this framing activity can of course affect how things get done. Neil again:

The constraints on funding have been that [it] tends to be for visible actions on the ground. The funders expect physical outcomes for their money, whether it's the number of trees planted or whatever. You know, there needs to be some sort of quantifiable output with a lot of funding. And I think that has meant that we've generally employed staff who have been very focused on the delivery of whatever piece of work it actually is. And what we've not been able to do is to actually develop an infrastructure that actually supports those project officers.

So there are costs or constraints when alliances are made. Things become more directed, they start to take a certain shape. Concrete to Coriander becomes concrete as matters are made to count. This links to the third problem with any naïve view of civic action.

3 Actions are co-ordinated, or made to link together
Central government, while certainly starting to pay lip service to the articulation of local needs, is also concerned to develop policies that apply to

the nation-state. Localities are collected together to form a population, for which a vast army of statistics on green space, carbon sequestration, health, and so on are produced and used to promote a national and increasingly international natural environment. The latter agglomerations are far from innocent guides to the making of natures. They are not the apolitical matters of fact that can be used to judge local schemes – they are themselves materially heterogeneous matters that need to be understood as outcomes of associations and political processes rather than the starting points for such schemas.

This has effects too. Here is one of the managers at 'The New Opportunities Fund' (now part of the Big Fund), a national lottery distributor.

> In terms of ministerial expectations ... when we get Parliamentary questions they'll tend to be, 'How many playing fields have you funded or such like?' ... and that's not necessarily what we've wanted to focus on but, I guess that's what we've increasingly been pushed towards focusing on ... We don't get asked about the community involvement and whether people are particularly happy about certain things having happened.

Numbers and neat objects like football pitches and trees circulate better and can be aggregated more easily than happiness and community involvement (the point has been made some time ago by, among others, Latour, 1987). In the process, complex matters are turned into translatable and calculable objects. Or perhaps more tellingly, these figured ontologies, a term Helen Verran (2001) uses to convey this making of matters, already form part of the way that situations turn up in the first place. The world is prefigured so to speak, and the practices of making things count tend to reinforce some enumerations over others. In this case, and as Callon and Law (2005) suggest, it might be less a case of rallying against calculation as a single practice, and more an investigation of what gets calculated and how calculations are made in different settings that is at issue.

Again, it is useful to highlight the times, spaces and materialities that are being enacted in this figured ontology. Relatively discrete objects are made and time is punctuated or limited to a set period. The returns must be 'in' within the funding agencies' time frame (in this case, three years). In gardening terms, things might be best kept simple. Much more than the bark-mulch-and-box landscape beloved of municipal authorities (see Figure 10.2 and also Chapter 8) might prove to be beyond the scope of the project frame.

One story to tell at this point is that calculative agencies (the application forms and the audit trails, the people, plans, and so on) produce effects. Certain things gain in significance and other things can drop off the register. For Neil, it is all the work of running an organization, of getting projects off the ground, of dealing with others, that is rendered less visible. It is the lack of core funding or unrestricted income that is the problem, the squeezing of overheads, the stuff that helps him build and repair infrastructures.

Figure 10.2 Municipal garden with the favourite material form of bark mulch and box (Brindley Place, Birmingham)

In its place we have restricted or project funds, specified and geared to deliver set returns. This is a common story. In conservation settings, for example, there are complaints about only ever receiving money to plant trees. It is more difficult to find funding for more complex ecological work. Trees are easily enumerated, made present, and fit into various climate-friendly, forest-friendly and other aspects of an increasingly target-friendly environment. What emerges, then, is a story of centralization wherein organizations and materialities are shackled and allowed little if any room for manoeuvre. Despite government rhetoric which emphasizes local action, things end up being centrally organized, micro-managed and orchestrated. Resources are on a tight rein and there is very little opportunity for creating bio-cultural diversities. There is a logic or discourse of calculation, governmentality and rationality that draws everyone and everything into its imperious gaze and order.

This is a neat, even satisfying story to tell. But in practice and politically, things are not so easy. Indeed, there are two responses to this kind of 'iron cage' story. The first relates to the empirical situation which doesn't quite fit this tale of centralization. The second relates to a political situation in which this kind of critical story tends to flatter to deceive, or doesn't quite get us as far as we might want to go.

First, Neil talks about a *game* that is played (see the earlier quotation). He hints that things might not be so set in stone. While the funding process is problematic, it is also malleable. There is a toing and froing, a driftwork

(Lyotard, 1994) perhaps, between actions and their co-ordination. The garden doesn't spring up from the grass roots, so to speak, but there is more to the process of environmental action and gardening than the delivery of central policy. In practice, neither voluntarism nor control, neither agency nor structure will suffice.

Second, even though there is an initial satisfaction in telling a critical story, the effect can be to overplay the power of the centre (see Box 10.1). Blaming the bureaucratic imposition of calculation, or the economy, or the market, or the government, the state or some other body can both over-simplify and over objectify the matters of which they speak. Gibson-Graham (1996) calls this type of story a romanticism of defeat, for in telling just how pernicious these things are, we risk making them out to be a lot tougher, more coherent and intractable than they may actually be *in practice*. As an important aside, it is worthy of note that critical stories can just as easily turn things on their head, complaining in this instance when there are no targets set, when records are not kept. It is impossible, we might say, to hold people responsible or make them deliver their promised returns if there is no practical effort to account for activities. So there might be a better politics to engage in than one that either castigates or unreservedly celebrates a specific set of practices.

Box 10.1 On being critical

A critical stance is one that is very common in social sciences and can be extremely useful. In this case it is to say that rationality, calculation and administrative bureaucracy are formatting the world in damaging ways, ways that are not recognizably ecological or relational, but are obsessed with neat objects, dead matter and equal, commensurate exchange. Ecological thinking and environmentalism have often taken this route, drawing on the Frankfurt School and critical theory more generally to denounce the modern calculating machine. The argument runs that the world is rapidly reduced to a spreadsheet of returns, a calculable mass, enframed as standing reserve (or *gestell* as Heidegger termed it) (Heidegger, 1978). The danger is that we end up making rationality and calculation sound more pervasive and more organized than they are in practice.

Here I follow another path, one that closely mirrors that taken by Annemarie Mol in her engagement with medical practice (Mol, 2002). Mol's ethnography of medical practice allows her to do something other than reproduce a critical medical sociology. The latter has long argued that patients' feelings, pain, meanings, lives and subjectivities (which medical sociologists call illness) are rendered unimportant or silenced by medicine's rush to the biomedical facts

(Continued)

(the disease). Disease can be thought of as nature to illnesses' culture. And yet, in her ethnography of clinical practices, Mol finds that patients are anything but silent, and that pain, feelings, lifestyles, and so on are as important as machines, measurements and pathology results in deciding what to do and what treatment to follow. In other words, in practice, illness is not displaced by disease. Both interfere with one another to produce a practical and sometimes non-coherent set of matters which are used to make treatment decisions. Instead of disease objects being used to silence ill subjects, an assemblage of subjects and objects is carefully produced as a means to making important decisions over how a patient is to go on.

In saying that subjects *are* excluded from medicine, as medical sociology tends to suggest, we run the risk of accepting that there *is* a neat divide between subjects and objects, illness and disease, feelings and pathology. But in Mol's ethnography we learn that the divides are fragile, and enacted. The result is that instead of offering another critique of biomedicine, accusing it of missing all the soft stuff, and being falsely founded on medical facts or dead matter from the pathology lab, Mol's field notes alert us to the rich variety of medical practice. In practice, clinical consultations are just as foundational as pathology results. So instead of escaping the biomedical model with criticism, which ends up building the model up to be more than it is, Mol advises that 'a good way to escape from a medicine founded on pathology [is] to wonder whether, in practice, medicine *is* indeed founded on pathology' (2002: 47).

Mol is not alone is taking this route. Foucault's work on sexuality is similarly concerned to avoid making the received categories of sexual identity beyond question (Foucault, 1981). Likewise, Callon and colleagues unsettle the purity of calculation in order to demonstrate its non-foundational character (Callon, 1998b; 1998c; Callon and Law, 2005).

The broader point is that in place of a criticism which makes the categories more solid than they need to be, we can engage practices to demonstrate that matters are more mixed up, more heterogeneous, than any pure logic or order would have us believe. The effect can be to destabilize those stories (critical and celebratory) that would have us believe that there is one way of doing things. Once destabilized, we can open up the practices to further discussion and experimentation, to more ontological politics.

So how to escape this story of the inevitable capture of good intentions by the administrative, surveillant machine? The question is similar to one that has circulated in social theory in response to conventional readings of Michel Foucault's earlier works. Foucault indeed tended to write of discourses as all-consuming orders, as structures that were imposed on a disordered world. He talked of the order of things, the birth of modern discourses of medicine and of surveillance and discipline (Foucault, 1970; 1973; 1977). The danger is, again, that in naming them thus, discourse becomes overly coherent and as impossible to unseat as the a-historical logics that it was trying to displace. So how can we do better?

Again Mol (2002) proves to be a useful guide. She highlights two strategies for working in a register that refuses the targets of conventional criticism. The first is to follow Latour in suggesting that rather than there being logical and coherent structures that shape the social, and that impose themselves on social settings, things and actions take shape in practice and are thereby liable to differ from any starting trajectory. There will be swerve (Latour, 1999). So rather than the garden being forced to work in a particular way as a result of an invisible and all-powerful logic or discourse (in this case perhaps of governmentality, or even the market), Latour would suggest that such logics or discourses have no power to impose themselves. Rather, their power is through association and their coherence is achieved through practical and material arrangements not through logic (the example of pasteurization proceeding through associations rather than through scientific logic remains a wonderful exemplar of this strategy, see Chapters 1 and 7 and Latour, 1988). The advantage of foregrounding practices of association is that the analytical playing field is levelled so that size and power become effects – no more and no less. Power becomes less certain, less totalizing. Contingencies threaten even the most established of entities or procedures. Everything is heterogeneous and thereby impure. Coherence is an effect, not a god-given or rationality-given condition.

While Latour's associations, with all their heterogeneities, their things and their tendencies to fail, are certainly more contingent than any discourse or logic, they can still retain a degree of orderly coherence. As many have argued (Lee and Brown, 1994; Mol, 2002; Munro, 1997; Murdoch, 2005); there is an homogeneity to Latour's associations. Things or elements are either associated or not, and are thereby either inside or outside the collective. There is, in these readings of Latour, little space for difference or for partial connections (see also Chapter 8). Different forms of relation and co-ordination tend to be effaced. Different modes of association and power are underplayed (Allen, 2003). Different modes of ordering (Law, 1994) are given short shrift and inter-network relations are given little airing (Murdoch, 2005). Whether or not Latour's work is rightly accused (there is a more complex geography than inside and outside in Latour, 2004b, for example), the point remains that it is important not to seek to replace homogeneity and order by what effectively amounts to more of the same.

This brings us to a second strategy for working in this alternative register. It is to argue for multiplication, for heterogeneity, and for more than one kind of organizing. As Mol helpfully puts it, many of these attempts to go beyond Foucault not only follow Latour in raising questions about the *force of* ordering, they also go on to raise questions about the *extent of* ordering.

So, if we return to the garden, what I want to hint at here is that these and other workings of time, space, objects and materialities are not following an inexorable logic that leads to ruin once frames and calculations are made. The world in other words is never so neat – there are matters that don't fit, surprises, room for manoeuvre, drifts as well as works. If we follow the practices further, we might start to find ways to unsettle the notion that calculation is all-powerful and is anti-ecological. We might find another escape from the rational planning model. The escape is to show that the rational planning/calculative model is not so pure, not so foundational, and is one of a number of modes of ordering that are in play. In practice, the funding agency model is heterogeneous, made up of many things and different modes of ordering all of which relate to one another in varying ways.

Modes of ordering

One principle of this book is that to contribute to the making of some thing, be it a microbe or a garden, is to both relate and to withdraw. It is to make a difference through and in relation – the seemingly paradoxical claim that the more that something becomes some *thing*, the more it is entangled with others. As we have seen in previous chapters, the more activity there is from laboratories, the more there is activity from their things. In other words, the relationship is not a zero sum game. Things make one another. This idea has been around for a while now in science and technology studies at least (the exemplar is usually Latour and Woolgar, 1979). Laboratories raise worlds. They secrete realties (for a neat retrospective, see Mol, 2002: 43). Gardens are good things to demonstrate this too – the more a garden takes shape, the more entangled it becomes with gardeners, who are of course not only human (in the simplest of lists, insects, micro-organisms, wind, catalogues, fertilizers, and so on garden the garden). Following this 'to be is to be related' move, another move is to say that not only do things take shape in relation, but they take *shapes*. That is, in becoming more real things also become multiple (again Mol, 2002 provides the clearest exemplar). For if we accept that things are done through practices, and that practices are heterogeneous, involving different places, people and others, then it follows that things will not be entirely settled matters. They will be pushed and pulled by a variety practices. So a garden will be the more or less coherent outcome of application forms, audits, drawings, voluntary labour, tools, plants, weather, gardeners, insects, and so on. How all these elements come together, in concert, in conflict or perhaps indifferently, is a matter that will also vary wildly, and is therefore an empirical question.

> The first months of the project involved the project officer meeting women, gaining their interest and support, and meeting their families. Early gardening work involved women practising and learning new gardening skills in back and front gardens, and in container pots. Small successes with the first easy-to-grow plants built confidence and generated interest in cultivation. Tools were purchased and a tool loan scheme established. The project worker spent large amounts of her time moving from household to household, keeping potential gardeners involved by circulating tools, supplying seeds, compost and other materials. Tools changed hands, and circulated. Plants and gardens changed form. Even hands changed as the women became handier at gardening. Training sessions were organized, allowing the women to develop new skills (not only in gardening, but in food preparation, healthy eating, and so on). The range of gardening activities increased and within the first year the scheme had moved from back gardens to an ambitious attempt to turn a derelict area of former parkland into a productive vegetable garden.

The gardens and gardeners make each other (this is the first point – they are made through relations). Work did not stop there – indeed, as the group developed and the gardens grew, the stakes increased. Maintaining relations is not easy. To get a sense of this, here is a typical day (derived from a research diary entry).

> Meeting at the CSV Environment Offices in Small Heath between 8.30 and 9.00, waiting for the project officer to arrive, finding as many garden tools as we can, checking someone has brought along the required organic compost, buckets of chicken-manure fertilizer, seeds, canes and any other supplies or equipment necessary for the morning's work. The equipment and materials are often in several different places (the project is a multi-site affair, making organization more difficult). So phone calls are necessary to other volunteers and to the gardeners to see who has what. Then, after a few more phone calls, we drive to several of the women's houses or flats to pick them up (stopping for a cup of tea and being introduced to family members). At one of the women's flats we fill at least thirty plastic bottles with water. (There is

no mains water supply at the 'derelict park' cultivation site, making the transporting of water across the city a weekly chore.) The woman who lives here has been helped by the project worker to move in and decorate the flat – her family was recently re-housed by the council and might well have been moved to another part of the city, which would have been a disaster for the project as she is a key member and motivator of the group. We then arrange to pick up more women. They all live within a mile or two of the sites but transport has become a major organizational issue. Most of the women are not used to or keen on walking this sort of distance and few had access to a car. Buses are not always convenient, particularly when carrying seedlings, gardening tools or water, and some of the women are not confident in using public transport. Fitting a morning's gardening in with otherwise busy domestic and work schedules means that the project officer is concerned to make transport to the site as easy for the women as possible. Finally, we weave our way through the Birmingham traffic to arrive at the site by (if we were lucky) 10.00. It remains only for the women to organize themselves and us so that the tasks to be done are completed by the time that many of them have to leave (around noon). Some of the women work in the polytunnel, thinning seedlings of plants that, they tell us, remind them of their youth in Bangladesh. Others clear plots of weeds and plants that have gone to seed. We help to dig the plots over, add fertilizer and sort out the blockage in the water butt ...

There is clearly more to the project than the framed returns that circulate on forms and spreadsheets. In *addition*, then, to the calculations and frames, there are muddles, problems to solve (involving water, transport, housing, language), instances of making do, pests and weeds, personal and family crises, childhood memories, carrots, tomatoes, rhubarb, familiar and not so familiar smells, aching limbs, elation at seeing the first shoots appear and at tasting garden produce. So the garden and the women's group become more real as they become more entangled. And, the garden is made up of multiple practices and is being pulled and shaped by many things. Clearly though, all these practices are not having equal effects, some are louder and more organized than others, some are almost invisible. Some contribute to the garden, some make life more difficult, some are made to co-exist with the garden and gardening, some need to be solved in order to keep the garden going. In order to understand more of this complex making of a garden, and to start to think about the effects that their ordering will have on the possible future of the garden, it is useful to think through the device of 'modes of ordering' (see Box 10.2).

Box 10.2   Modes of ordering

In *Organizing Modernity,* John Law develops a set of four 'performative' 'little narratives' or 'modes of ordering' which are both told and embodied in the non-verbal practices and materials of an organization (Law, 1994: 20). Briefly, modes of ordering have the following characteristics (see also Chapter 7). First, in any setting there will be more than one mode of ordering at work. Second, they often depend on one another for their existence, but may be in competition, may go on despite one another, and so on. In other words, they often interact or relate to one another in ways that can assist each other or threaten the other's existence. Third, they mark *attempts at* ordering rather than orders. They are not logics or rationalities visited on the scene from outside, but endeavours to enact something that are bound, through their socio-material heterogeneity, to perform themselves imperfectly. Fourth, and following this, modes of ordering simultaneously make actors and contexts, agents and organizations. Neither comes before the other. So modes of ordering do not have a thinker or actor at their centre, they are non-anthropocentric. Modes of ordering make many things, including in some circumstances actors and things, individuals and organizations. Fifth, and finally, they are devices that are crafted from ethnography and fieldwork, and in that sense are a product of field site and fieldwork. They are both out there, being done, but also categories devised by a fieldworker to sense and intervene in those practices.

Law makes the argument that far from being general types, such modes are contingent, developing as peculiar interventions in particular settings and also, of course, are formatted and made sense of in particular ethnographic settings. For Law, doing his ethnography in a large laboratory in the early 1990s, the modes of ordering included enterprise, administration, vision and vocation. This was a particular setting, one where science was increasingly asked to pay its way and account for itself.

We wouldn't expect the same modes of ordering that Law identified in a laboratory to be useful in understanding charity funding and urban gardening settings. For as Law is at pains to insist, modes of ordering are contingent, empirical matters, and diversity and change would be expected (Law, 1994: 83). In the story so far, however, and perhaps in keeping with the ways in which organizational arrangements have circulated in the UK for a number of years, there are certainly recognizable elements of administration and enterprise modes at work in the charity sector. Ensuring that

TABLE 10.1   SUMMARY OF MODES OF ORDERING IN URBAN
ENVIRONMENTAL ACTIONS

Mode of ordering	Comment
Administration	Record keeping, audits, forms
Enterprise	Playing the funding game, opportunity funding
Remediation	Repair, one-off funding, action-led, making things present, enabling and rehabilitation
Care	Engaging, longer-term commitments, patience, maintaining relations

records are kept, procedures are followed, and things are made accountable tells of an administrative mode of ordering. Meanwhile, charity funding, projects, organizations – all tend to be enacted in ways that tell of opportunism and performance and thereby enterprise. These administrative and enterprise modes of ordering interact with one another, and make one another. So, for example, the director of CSV Environment, somewhat reluctantly, plays the funders' game in ways that are surely entrepreneurial. Playing the funding game is about finding ways of using the bureaucracy to carve out opportunities for doing the 'real work' of environmental actions.

Two other modes of ordering are worthy of attention here (see Table 10.1). They both enact urban restoration and regeneration. The third we can call remediation. It's the attempt to provide a quick solution for a problem or area, often through short, fixed period funding which involves an injection of money in order to 'kick start' an activity. The medical language is informative, for the approach is similar to that of a patient with an acute condition that can be cured with drugs. Funding environmental action in Small Health and Saltley would provide a partial cure for a number of problems, including cardio-vascular complaints, lack of accessible open space, isolation, and so on. Remediation tends toward a short course of funding, designed to get a patient back on her feet. It speaks of enablement.

> One of the things about lottery funding is that it is supposed to be short-term funding, sort of kick starting things … We do always tend to be looking at new projects to be funded. We wouldn't necessarily be able to fund a project that had already been going for some time. (Manager, New Opportunities Fund)

The fourth mode also has a medical provenance (perhaps not so surprising given the history of charitable activities). But rather than the quick administration of a cure, here the mode of ordering is longer term. The approach has more in common with the treatment of chronic problems, where attention is turned not so much to curing the patient but to developing a caring relationship (although the divisions are hardly clear-cut in practice, as Pols, 2003, shows in a study of modes of caring and their relation to changes in juridical orderings of patients). Instead of enabling, as such, it speaks of engaging (see Box 10.3). In the current case, it can loosely convey something of the relationships that are generated between the women and the project officer, between the women and other group members, between CSV Environment and this part of the city, between the women and the plants and between the garden and the group. Care is involved in the watering of the garden, in the pinching out of shoots, in the project officer's interventions to make sure members have good places in which to live. But it's not only involved in these one to-one relations between people and between people and plants. Care is taken in the filling in of forms, wherein categories are played with in order to not get too bogged down in the numbers game. It is involved in CSV Environment's desire to stay in this part of Birmingham even when the funding wave has moved on and circumstances suggest that it would be easier to develop new projects elsewhere. In these and other actions, care is certainly not the only or even the dominant mode of ordering, and nor is it one type of thing or addressed to one kind of thing or object, but it is there, in the mix of activities that make the garden grow.

Box 10.3   Notes on care

Care is a complex term and one that can have quite conservative connotations, especially perhaps when it is linked unproblematically to an ethic of care which assumes what Massey (2004) calls a Russian doll geography, one where care for the close at hand always comes before extending care for others. But it can evoke something less fixed, something that is not as controlling or spatially limited as implied by an extension of intimate relations (Benhabib, 1986) or the formation of an ethics of solidarity (Fraser, 1986) (see previous two chapters for hints at how this might work for human–nonhuman relations). In this more open sense, care is not nurturing in order that those cared for become the same as the carer. Rather, care here is linked to a curiosity, to a willingness to shift for and to others, an openness to others, a listening and a patience (for a clear statement, see White, 1991).

It is worth repeating the points made already in Chapter 8 that, contrary to the more humanist engagements in thinking about care (including White, 1991),

> *(Continued)*
>
> care need not be conceived as a virtue, a trait that we can somehow cultivate in ourselves, figured as lone bodily projects (and so perhaps departs from some of the practical philosophies of Foucault, Rorty and others (Shusterman, 1997)). Rather, care is produced with and as others, and is neither selfless nor only about the self. Indeed, it is an ecology that is not oriented to securing an inside (an us) nor oriented to everything outside, but a gathering together that is not too tight and can thereby work to confirm rather than to assimilate others.

The more general point is that there is more than one mode of ordering necessary to make the garden grow. Enterprise without administration would soon peter out. Remediation and administration seem to go hand in hand, but both seem to require entrepreneurial modes of ordering to get anything done. Meanwhile, administration without care would soon have the women not turning up for a morning's work. It is important to note that I haven't suggested that these modes of ordering occur exclusively in separate places (for example, administration in the lottery offices, enterprise at the CSV Environment offices, remediation in London and care in the garden). I don't want to suggest an easy geography of ordering. Rather, the modes are assembled in all these places with varying effects and emphases. The questions now become; how do these various modes interact, and how might their mixings, the ecologies of action, affect the gardens? How, in other words, is this multiplicity of gardens assembled? And, can it be assembled in better ways?

Ecologies of action

One common story in the social sciences is to claim that calculation or some version of it (like rationality) eventually out-competes or destroys other forms of life. The world becomes disenchanted, as Weber among many others would have it (see Bennett, 2001; Weber, 1991). But within the science studies tradition, rationality and calculation are not the inexorable, general logics that are marching to erase diverse, particular ways of being. Instead they are effects that have to be worked at and produced. Just as science studies scholars managed to demonstrate that truth was a complex achievement of laboratory practices, so rationality becomes, in the hands of Callon and colleagues, a matter of heterogeneous practice (Callon, 1998c). Moreover, in making calculation a practice, there is room for political manoeuvre. Matters are not pre-set, but open to variation and to intervention. The extension to this argument is that once we note that calculations are practical achievements, somewhat messy and heterogeneous, and once we note that modes of ordering are themselves heterogeneous and insufficient on their own, then attention can be turned to the mixtures of activities, or ecologies of action, that contribute to a state of affairs. In any empirical situation, then, we can ask how ecologies of action can be improved.

> **The Concrete to Coriander project has enabled all manner of things to happen in east Birmingham. The project is a success in terms of the project frames and the returns. The raised beds are productive, the women talk of their enjoyment and desire to carry on meeting. Photographs tell a similar story, of dereliction to cultivation, and are used as evidence of success that can circulate between field and office (see Figure 10.3). Gardens have been made all over this part of the city. Some things have been made concrete, disentangled and delivered to the satisfaction of funding agencies and steering groups.**

So the project is a success. CSV Environment, like the funding agency, can walk away. The returns are in. And yet, as project money came to an end, things started to look less solid or more and more unfinished, embroiled, entangled. There were unresolved problems, like transport and water, and a strong sense that without support the women would find it tough to continue their activity. And of course the returns that could be disentangled and made to stand as a measure of success could only work on account of the complex of entanglements that made them possible (the women's reduced isolation is in large measure a result of greater ties, etc.). So things were far from being neat – nor should they be, for gardens and groups need entanglements to continue. In this gardens are if not unique, then particularly problematic. Gardens are matters of process, and ongoing achievements. Once created, gardens require continuing work. So how to carry on once the returns are all in?

Here is what is supposed to happen. All the funding agencies should be able to walk away, for the funds invested should have created the opportunity for self-sufficiency and sustainability (which means in this instance, an ability to continue without further outside funds). The remedial funding should have created networks, reciprocities and associations which can form resources for mutual benefit. The term that is often used for this social sustainability model is 'social capital'. The terminology is from social science, and most notably from the work of sociologist Robert Putnam (though with important antecedents in Jane Jacobs and Pierre Bourdieu; Bourdieu, 1986; Putnam, 2000). Social capital has moved through policy circles like wildfire (Fine, 2002). It speaks of social (people to people)

Figure 10.3 Before and after photos of the derelict park site. Pictures like these are part of the 'return' to funding agencies, telling the story of the project.

bonds, treating those bonds as assets that can provide a stream of utility to individuals. Thus funding agencies seek to format and frame projects that will deliver social capital. Any requirement to carry on funding is therefore regarded as a failure, for insufficient capital has been made. Remediation has failed.

To be sure the women are less isolated, they are more likely to help one another, they have met and engaged with other allotment gardeners. They have formed a co-operative which shares tools, seeds and vegetables, and they have learned and practised skills and gained qualifications in food preparation. The women now provide cooked food at numerous public fetes, environmental days, school and community events, and so on. There is even talk of running a café, using produce from the gardens. But many of the women are also relatively elderly and close to retirement age. They have numerous domestic and work commitments and, for some, are reliant on social welfare payments (which might be jeopardised if the group developed their economic activity in particular ways). There's a feeling that the group and the gardens are far from being self-sufficient. More time and more money are needed to keep things going.

The short version that is Concrete to Coriander cannot be a success on these social capitalist terms. Indeed, as the campaign group, Black Environmental Network, has put it, many like it cannot succeed on these terms:

> Funders tend to consider their commitment in the short term and challenge small organisations with providing an exit strategy which would make no further demand on the funder as proof of their being fit to have a grant. This is experienced as an enormous and unfair burden, especially when a high proportion of projects from ethnic groups are ones which enable ethnic groups to get a first foothold on the problems to be addressed. (BEN, 2000: 31).

It is worthwhile pausing here to ask what is being enacted when social capital becomes the objective of funding agencies. The answer, in the Concrete to Coriander case, is three-fold.

1. It enacts a single, self-identical capitalist economy to which people are more or less successfully connected. Areas low in social capital are by definition poorly equipped with respect to the economy. Boosting associations of this type will, it is argued, enable people to develop economically. Any failure to develop self-sufficiency is a failure of the project. There is no recognition of other kinds of economies, from co-operatives to domestic labour, which are implicated in and can interfere with a not so homogeneous capitalism (see Gibson-Graham, 1996).
2. It enacts a particular ecology of action whereby some modes of ordering are championed over others. Remediation, and thereby one-off funding

followed by self-help, are privileged over longer-term care. Administration and thereby governance are privileged over enterprise. Although, as we have seen, nothing much would happen if this privileging turns into dominance, the danger is that the ecology of action associated with social capital results in modes of ordering being set against one another, so that they clash rather than work in tandem.

3. Finally, it enacts a particular version of the social, one where people are treated independently to other things. The result is that if the gardens close or become disused, then it is of no real concern, for what is central is the development and transfer of social capital. Human relations, or particular kinds of human relations, are foregrounded. As long as the women move on, and develop as economic agents, then the gardens become immaterial to the sustainable development of east Birmingham.

Another way of saying all of this is that the current arrangement tends towards the enaction of singularity rather than multiplicity, a state of affairs that I will argue is a long way from environmental sustainability. I want therefore to finish this chapter by revisiting what sustainability can become in view of both of the case study presented in this chapter and the arguments that I have been making regarding geographies of nature.

Sustainability and multiplicity

Two versions of sustainability have appeared so far in this book. In Chapter 8, there was a standard, prudent, environmentalist account whereby sustainable living involved careful planning, taking care not to use resources unnecessarily. The histories of this prudence are long and far from straightforward (for Western-centred approaches, see Glacken, 1967; Passmore, 1980), but one significant episode has been the development of North American conservation and preservation arguments, taking in the gospel of efficiency and the wise use movements (Hays, 1959). There's a lot to recommend in this non-profligate approach to living, particularly when the consequences caused and rewards gained by massive and geographically uneven throughputs of materials and energies are so overwhelming and underwhelming in turn. But I also hope that by now it is clear that while such matters are hardly unimportant, there is more to 'taking care' and to sustainability than this prudent model conveys. One argument of this book has been that conservative approaches to nature tend to assume a position of human control, a passive nature and a compulsion to infinitude (to rendering the present eternal) (Bowker, 2004). In short, in the history of ideas about conservation and preservation, nature is treated as a singularity, as a mass to be used more or less successfully. In contrast, I have tried to demonstrate that *in practice* natures are far from controlled, determinate or self-identical. Instead they are practised by all manner of things and are thereby multiple. To practise sustainability is a multiple exercise.

The second version of sustainability, which appeared in this chapter, is even more suspect. Instead of a fixed and independent nature, we have a sustainability of no nature, or a sustainability where nature is but a backdrop which does not enter into the machinations of human political economy. This is sustainability where the social is stripped of all the matters that make it even remotely possible in the first place. It's the sustainability of neoclassical economics, of the sociology of the social as Latour calls it (Latour, 2005), of the social capitalists, and so on. Sustainability here means economically viable, able to sustain itself, to be monetarily self-sufficient. As I have suggested, it treats 'economy' as singular. It also treats nature as singular, as one thing which can be happily placed in the background. In treating nature as singular, both versions of sustainability share the same result even though one attaches more importance to that nature (see Chapter 3 for a discussion of the similarities in this sense of nature-independent and nature-dependent stories in treating nature as an unchanging single mass).

How can things be assembled in ways that are more sustainable? The short answer is to pay attention to the ecologies of action, to the interrelations that exist within and between the multiple practices, modes of ordering and materialities that go to make a garden. To be sure, there is no magic formula or balance to be struck, producing good ecologies of action can only be experimental, but the salutary lesson is that any attempt to assemble a garden, landscape, city, policy or other grouping will require more than one mode of ordering and a sensibility that is open to those matters and practices that have been temporarily obscured by the requirement to make things happen and to generate coordinated actions. If there is a mantra, then it would be to make sure any action or closure is accompanied by attention to what is outside the current set-up, and how various members of a gathering are only ever partially given to that set-up. It is a loose kind of knowing and assembly that makes for more progressive and sustainable ecologies of action. In terms of the case study, funding agencies vary in their approaches, and this variation opens up possibilities for doing things differently. However, there are moves in lottery and government regeneration funding towards tight coordination and close specification. The result may well be an ecology of action that is indifferent to the complexities and multiplicities involved in making things that matter. Finally, in pointing to a caring mode of ordering I have suggested, both in this chapter and in previous chapters, that making spaces for nature involves a stance that allows others to flourish. To be sure, in the practical politics of making things happen, care is hardly sufficient. There are requirements to make things present, to get things done, to establish routines and accountabilities, to get money out of the door and to trace its use. But without the generative geographies of care, with their willingness to patiently allow others to emerge and to differ, policies and actions risk eradicating the very things that make things matter, their multiplicity.

Conclusion

Environmental actions don't just happen, they are coordinated, formatted, framed, made accountable, and so on. One way of talking about these routines and procedures is to say that they are anti-ecological and that they detract from the real business of making biodiversities. In this chapter I have argued that while there is plenty that is anti-ecological in policies and programmes, we may be missing the target if we label all forms of calculation and related organization work as the problem. Rather, it is more productive to investigate the heterogeneity of actions and in doing so to argue against those tendencies to purify or obscure the richness that makes action possible. Here I have used the notion of modes of ordering to say that there is no administration without enterprise and no remediation without care, the point being to argue that rather than administration or bureaucracy being *the* problem, it is the ways in which such matters are done, or are enacted along with other modes, that pull and shape things like gardens into particular forms. The politics to pursue here is then not one of blanket criticism but an ontological politics that involves itself in the making of realities. Finally, in introducing the term ecology of action, I have suggested that the ways in which the multiple enactions of things, like gardens, are assembled have effects on the interrelations that exist between the multiple modes of action. Attempts to enact a singular version of economy or nature can have the effect of making some modes of ordering, like care, less visible. Doing sustainability as a multiple would, in contrast, involve ecologies of action which create spaces for nature, for a gathering together that is open to difference, a non-coherent assembling.

Further reading

The ideas in this chapter can be followed in a good deal more depth in various places. A crucial read on the multiple is Annemarie Mol's (Mol, 2002) *The Body Multiple*. John Law's *Organizing Modernity* is a formative text here too, and his work on foot and mouth also builds on this approach to the multiple doings of the social (Law, 1994; 2006a). For more on generosity and care, Ros Diprose's *Corporeal Generosity* (2002) develops a sophisticated attempt to avoid reducing interrelations and others to exchange and sameness.

Afterword

Activating Geographies of Nature

The natures that animate the pages of this book are different. They do not obey the easy cuts of Nature and Society, Science and Politics, to which many have become accustomed. Indeed, the old divides are counterproductive. On farms and food processing plants, nuclear waste repositories and military battlefields, hospital wards and government offices, the practices of making some thing happen are being let down by the anti-politics of Nature. Matters are closed down too quickly, and with too great a degree of irrevocability, with the result that reality comes back to haunt those who have declared things to be safe, under control, or no longer worthy of attention. It is increasingly clear, then, that the space-times of things are not well served by the 'one stop shop' that Nature offers.

Similarly, Nature doesn't seem to be working as a rallying site for everyone and everything anymore. If in the past it has allowed us to terminate debate and – like God before it – secure an agreeable collective, things certainly don't seem to be as easy now. Already, faced with an unruly and heterogeneous populace, a massive demographic of humans and nonhumans who won't fall into place (Callon and Law, 2004), what or who counts is very much up for grabs.

How do we proceed politically without Nature to ground us? How can a democracy be built that refuses to kowtow to Nature, but at the same time takes into account the space-times of the new demographic? There are at least two possible general means of moving these questions forward. The first is to proclaim 'no nature', to work towards political settlements in a purely cultural register. The second is to argue for many natures, a kind of multi-naturalism (de Castro, 2000; Latour, 2003).

No Nature, or politics without Nature

Certainly the one nature, or the timeless space of nature that marks an attempt to settle disputes, to declare the truth of the matter, the matters of fact, as Latour (2004b) calls them, is both objectionable politically, and, it seems, is increasingly a rather weak myth by which to govern affairs. One response is to retreat from any notion of nature, to declare the ruse of nature to be well and truly finished and seek alternatives. If a singular nature never really existed, then we are left only with a social world, narrowly defined as consisting of human beings and their interrelations, within and upon which to develop a collective. And yet, every attempt to declare a purely social order merely re-invents a natural counterpart, and in so doing merely reproduces the old Kantian settlement of society/nature. Far from being a crude idealism, the lurch to the social tends to reproduce, as its mirror, a singular nature. The dynamic social world creates its other, a dead world of non-social matters, a world of timeless bare facts that can be called upon again to settle debates once and for all (Latour, 1999). Indeed, as Mol (2002) has demonstrated effectively, in focusing attention on the who rather than the what of politics, and despite setting out to democratize proceedings by problematizing who gets to decide, we end up with a residual nature that is both lacking in vitality (Fraser et al., 2005) and which produces the same old short cuts or deter-mined outcomes as before (see Chapter 3 and Hinchliffe, 2001).

To explain this further, the focus in this no nature politics becomes only one of discussing 'who' gets to decide what should happen and how to orga-nize the who so they can decide as 'freely' as possible. Organizing this 'poli-tics of who' takes at least two general forms, both suffering ultimately from a dualism which leaves Nature as a problematic, anti-political, residual. Constructing markets is one way of developing a 'politics of who', in this case, giving consumers the ability to direct resources. The mechanisms are varied, but the general theme is of finding ways of providing choices to indi-vidualized decision-makers, who may 'choose' through their purchases to col-lectively produce better environments, particular medical interventions, and so on. The second general form of democratization is of who is civic, where the focus is on more deliberative and collective formations, debating the best or least cost options. But even with their well-known differences, what market and civic methods share, as Mol (2002) says, is a suspicion of experts and professionals, those who in the modern constitutions would have deter-mined matters in the name of the external arbiter, in this case, in the name of nature. And yet, ironically perhaps, they also maintain or even underpin the notion of expertise. For in this social and moral world where humans debate and choose, experts are now asked only to give us the facts of the matter, the unbiased bare facts which can be laid out, transparently, before the citizenry or on the market stall.

Even if such ruses of bringing nature back in as bare fact are avoided, there are other problems with this 'politics of who'. One example would be the long-standing tendency to assume the issue for politics is simply one of giv-ing voice (to people with already formulated desires). Another would be the tendency to simplify choices as momentary decisions rather than practical achievements made up of the numerous 'intertwined histories that produce them' (Mol, 2002: 169). But as Mol and many others have noted, the subject and the we of politics is in no simple sense 'us'. As Haraway put it, 'all of the actors are not human and all of the humans are not "us" however defined' (1992: 67).

Given such concerns, Mol offers us a *politics of what*, a politics where expertise and the things of expertise are neither determining nor immaterial to finding ways forward. Moreover, it is a politics that makes use of, rather than patronizing, tolerating or ignoring (or most likely all three), differences and different ways of understanding and enacting things (be they a body, a disease, a climate, an environment ...). What Mol, Haraway and Latour share in this respect is a version of politics that takes who *and* what seriously, and seeks the means to refuse to settle debates on one side or the other. So this is a politics that is materially and socially multiple, and that attempts to find ways of moving toward assembling answers that refuse to eviscerate the politics of what by declaring matters to be natural or social.

Many natures – doing politics as multinaturalism

Even if we sympathize with the project of democratizing the sciences and/ or of articulating a politics of what, is a concept of nature really needed? Do natures still have work to do? This book has started from the premise that there is a case to be made for natures. There are a number of arguments which together suggest that the processes of reconstituting politics is one that can be enriched, rather than impoverished, by natures. Let us take each in turn.

First, the natures that are talked about here are not starting points. They are not the initial grounds for politics (or more likely anti-politics). Rather, if they arise they are the sometimes fragile, sometimes fairly robust, end points of a complex of activities. Microbes, rare species, diseases, animals, gardens are all matters that *in the end* are more real than representations, but whose reality is not outside the fraught constitution of political world-making (Latour, 1999). They are the outcomes of all manner of works and mark more or less stable assemblies which can themselves do further work. But in being produced, they are not necessarily determined by their relations. They are objects that have objected, that have become more real as the constitutional work has gathered apace. It is this apparent contradiction between a manufactured object that is nevertheless more than human that marks the interest in geographies of nature. Neither made up nor pre-existing, both formed and forming, natures mark the lures that mobilize an indeterminate world.

Second, if natures mark the irreducible and indeterminate outcomes of activities, then they can also figure in the opening up of apparently settled matters. Any closure, or attempt to settle matters, any constitution, will create outsiders, matters that are not known about, or if known, matters that are considered irrelevant to the business of going on. Many of these natures will happily work alongside an assembly, without a grumble. But then there are those that will demand to be taken into account, or those things that refuse to be ignored. They are the surprises, the unaccounted for, that produce political and scientific events that re-open the settled collectives (be they scientific norms or political institutions).

Third, geographies of nature are not solely about the openness of the world in terms of its refusal to be closed. This is more than a politics of insiders/ outsiders or of the process of othering and dissent that goes on every time a constitution is formed. *Geographies of Nature* also marks an attempt to consider what being given to the world can involve. 'Being given' is not synonymous with being already established, timeless or fixed. Being given marks the play between being open to others at the same time as making and marking a difference. There is then a generosity that inhabits geographies of nature, a concern with and for others, not only after the event, as they come back to haunt our schemes and assemblies, but in the very make-up of the world (Diprose, 2002). In short there's an ethos here that is generous with and to others. Which means that geographies of nature are complex, involving others (but not, it should be emphasized, suffocating otherness). In this, to return to an opening problematic, there is to be sure more than one nature. Natures are multiple. But this is not a statement of perspectival politics or even pluralism. Multinaturalism is not relativism. The politics here is an ontological process, subject to various modes and forms of power, as things are pulled and shaped by numerous practices in numerous places with numerous interrelations.

So what is to be done? This book has opened up the possibility for a geography of natures, and *for* nature rather than arguing for its end. It has also suggested that any doing of nature will be multiple, spatially and materially. The assemblage of nature is in process and the processes can be engaged in through many different activities, practices and places. How to engage in the making of better natures is a fraught empirical and political question. I have suggested that the question is both ontological and political, and requires detailed engagement in the multiple practices of nature making. If nature is done, in lots of ways, places and with lots of others, then rather than offering interpretations of nature, or analytical concepts, the injunction must be to join the doings, to experiment, to engage in the doing of environments, to environ in different and better ways (Thrift, 2005). The examples given here have only been indicative. I have not been able to elaborate on the roles that can be conjured for scientists and social scientists in this world-making. Big questions remain or need further experimentation. For example, once the old forms of criticism have been surrendered, how are social scientists in particular to operate? Ethnographies of natures and experiments that don't necessarily have human-being at their centre and which attempt to change the make-up of an assemblage are a starting point. Clearly though, any old experiment won't do, and there are codes to develop and normativities to build. While the old normative certainties have gone, as nature has moved from the past-present and another country to the future-present and to a multiple spatiality, a non-foundational framework for moving forward is starting to take shape (Latour, 2004a; Mol, 2002; Stengers, 2000), one that offers roles and places for social scientists to involve themselves. This involvement will not be to act as interpreters or as legislators with some hot line to the truth (Bauman, 1992), but as co-generators of more and different representation-interventions to an assembly (representations that are, it needs always to be emphasized, matters that are made rather than pure images of the world) (Latour, 1999; 2004b; Whatmore, 2003). It may be to act in tandem with some groups, it may be to work alongside others, and in opposition to many others, but it is always to generate more things, to add more to the world (rather than to subtract from it). The multiple geographies of nature suggest that there are many 'wheres' for doing nature politics, many sites and organizations to engage. This book has started to provide an opening to map some of those spaces in order that we can engage with matters that are not, after all, so dead and buried.

Bibliography

Adam, B. (1997) *Timescapes of Modernity: The Environment and Invisible Hazards*, London: Routledge.

Adams, W. M. (2004) *Against Extinction: The Story of Conservation*, London: Earthscan.

Adams, W. M. and Mulligan, M. (eds) (2003) *Decolonizing Nature: Strategies for Conservation in a Post-colonial Era*, London: Earthscan.

Agamben, G. (2002) *The Open: Man and Animal*, Stanford, CA: Stanford University Press.

Akrich, M. and Latour, B. (1992) 'A summary of a convenient vocabulary for the semiotics of human and nonhuman assemblies', in W. Bijker and J. Law (eds) *Shaping Technology/Building Society: Studies in Sociotechnological Change*, Cambridge, MA: MIT Press.

Allen, J. (2003) *Lost Geographies of Power*, Oxford: Blackwell.

Alpers, S. (1989) *The Art of Describing: Dutch Art in the Seventeenth Century*, London: Penguin.

Anderson, K. (1997) 'A walk on the wild side: a critical geography of domestication', *Progress in Human Geography* 21: 463–85.

Ansell Pearson, K. (1997) *Viroid Life: Perspectives on Nietzsche and the Transhuman Condition*, London: Routledge.

Ansell Pearson, K. (1999) *Germinal Life: The Difference and Repetition of Deleuze*, London: Routledge.

Atkinson, A. (1991) *Principles of Political Ecology*, London: Belhaven Press.

Babtie, G. (2001) *New Hospital Proposal: Water Vole Survey*, Glasgow: Babtie Multidisciplinary Consultants.

Baker, S. (2000) *The Postmodern Animal*, London: Reaktion.

Barker, G. (2000) *Ecological Recombination in Urban Areas*, Peterborough: English Nature.

Barnett, C. (2005) 'Ways of relating: hospitality and the acknowledgement of otherness', *Progress in Human Geography* 29(1): 5–21.

Barry, A. (2005) 'Pharmaceutical matters: the invention of informed materials', *Theory, Culture & Society* 22(1): 51–69.

Bauman, Z. (1992) *Intimations of Postmodernity*, London: Routledge.

BEN (2000) 'Funding issues affecting ethnic communities', Black Environment Network http://www.ben-network.org.uk/resources/downlds.html (accessed 22nd July 2004).

Benhabib, S. (1986) 'The generalized and the concrete Other: the Kohlberg-Gilligan controversy in feminist theory', *Praxis International* 5: 402–24.

Bennett, J. (2001) *The Enchantment of Modern Life: Attachments, Crossings and Ethics*, Princeton, NJ: Princeton University Press.

Bennett, J. (2004) 'The force of things: steps to an ecology of matter', *Political Theory* 32(3): 347–72.

Bensaude-Vincent, B. and Stengers, I. (1996) *A History of Chemistry*, Cambridge, MA: Harvard University Press.

Benton, T. (1989) 'Marxism and natural limits: an ecological critique and reconstruction', *New Left Review* 178: 51–86.

Beschta, R. L. (2003) 'Cottonwoods, elk, and wolves in the Lamar Valley of Yellowstone National Park', *Ecological Applications* 13(5): 1295–309.

Bingham, N. (2006) 'Bees, butterflies, and bacteria: biotechnology and the politics of nonhuman friendship', *Environment and Planning A* 38: 483–98.

Blythman, J. (2006) 'So who's really to blame for bird flu?' *The Guardian*.

Boundas, C. V. (1996) 'Deleuze-Bergson: an ontology of the Virtual', in P. Patton (ed.) *Deleuze: A Critical Reader*, Oxford: Blackwell.

Bourdieu, P. (1986) 'The forms of capital', in J. Richardson (ed.) *Handbook of Theory and Research for the Sociology of Education*, New York: Greenwood Press.

Bowker, G. (2000) 'Biodiversity datadiversity', *Social Studies of Science* 30: 643–683.

Bowker, G. (2004) 'Time, money and biodiversity', in A. Ong and S. J. Collier (eds) *Global Assemblages: Technology, Politics, and Ethics as Anthropological Problems*, Oxford: Blackwell.

Braun, B. (2005) 'Environmental issues: writing a more than human urban geography', *Progress in Human Geography* 29(5): 635–50.

Braun, B. (2007) 'Biopolitics and the molecularization of life', *Cultural Geographies*, 14: 6–28.

Bright, C. (1999) *Life out of Bounds: Bio–invasions in a Borderless World*, London: Earthscan.

Browne, J. (2003a) *Charles Darwin: Voyaging*, Vol. 1, London: Vintage.

Browne, J. (2003b) *Charles Darwin: The Power of Place*, Vol. 2, London: Vintage.

Budiansky, S. (1997) *The Covenant of the Wild: Why Animals Chose Domestication*, London: Phoenix.

Butler, J. (1997) *Excitable Speech: A Politics of the Performative*, New York: Routledge.

Callon, M. (1986) 'Some elements of a sociology of translation: domestication of the scallops and the fishermen of St Brieuc Bay', in J. Law (ed.) *Power, Action and Belief*, London: Routledge and Kegan Paul.

Callon, M. (1998a) 'An essay on framing and overflowing: economic externalities revisited by sociology', in M. Callon (ed.) *The Laws of the Markets*, Oxford and Keele: Blackwell and Sociological Review.

Callon, M. (1998b) 'Introduction: the embeddedness of economic markets in economics', in M. Callon (ed.) *The Laws of the Markets*, Oxford and Keele: Blackwell and Sociological Review.

Callon, M. (ed.) (1998c) *The Laws of the Markets*, Oxford and Keele: Blackwell and Sociological Review.

Callon, M. and Law, J. (1995) 'Agency and the hybrid collectif', *South Atlantic Quarterly* 94(2): 481–507.

Callon, M. and Law, J. (2004) 'Introduction: absence – presence, circulation, and encountering in complex space', *Environment and Planning D: Society and Space* 22(1): 3–11.

Callon, M. and Law, J. (2005) 'On qualculation, agency, and otherness', *Environment and Planning D: Society and Space* 23(5): 717–34.

Castree, N. (2004) 'Nature is dead! Long live nature!' *Environment and Planning A* 36: 191–4.

Castree, N. (2005) *Nature*, London: Routledge.

Clark, N. (2002) 'The demon seed: bioinvasion as the unsettling of environmental cosmopolitanism', *Theory, Culture Society* 19(1–2): 101–25.

Coetzee, J. M. (1998) *Boyhood: A Memoir*, London: Minerva.

Coetzee, J. M. (1999) *The Lives of Animals*, Princeton, NJ: Princeton University Press.

Coetzee, J. M. (2000) *Disgrace*, London: Vintage.

Collier, S. J. and Lakoff, A. (eds) (2006a) Distributed Preparedness: Space, security and citizenship in the United States, unpublished paper, available at http://www.anthropos-lab.net/.

Collier, S. J. and Lakoff, A. (2006b) 'Vital systems security': laboratory for the anthropology of the contemporary: http://www.anthropos–lab.net/publications/index. html.

Cronon, W. (1996a) 'The trouble with wilderness; or, getting back to the wrong nature', in W. Cronon (ed.) *Uncommon Ground*, New York: Norton.

Cronon, W. (ed.) (1996b) *Uncommon Ground*, New York: Norton.

Crosby, A. (1986) *Ecological Imperialism*, Cambridge: Cambridge University Press.

Cuomo, C. (2003) *The Philosopher Queen: Feminist Essays on War, Love, and Knowledge*, Lanham: Rowman and Littlefield.

Cussins, C. (1996) 'Ontological choreography: agency through objectification in infertility clinics', *Social Studies of Science* 26(3): 575–610.

Darwin, C. (1998) *The Origin of Species*, Oxford: Oxford University Press.

Daston, L. (ed.) (2004) *The Moral Authority of Nature*, Chicago: University of Chicago Press.

Davis, M. (2002) *Dead Cities*, New York: The New Press.

Davis, M. (2005) *The Monster at our Door: The Global Threat of Avian Flu*, New York: The New Press.

de Castro, E. V. (2000) 'Cosomological deixis and Amerindian perspectivism', *Journal of the Royal Anthropological Institute* 4: 469–88.

de Laet, M. and Mol, A. (2000) 'The Zimbabwe bush pump: mechanics of a fluid technology', *Social Studies of Science* 30: 225–63.

Deleuze, G. (1999) *Foucault*, London: Continuum.

Deleuze, G. and Guattari, F. (1988) *A Thousand Plateaus: Capitalism and Schizophrenia*, London: Athlone.

Deleuze, G. and Parnet, C. (1987) *Dialogues*, New York: Columbia University Press.

Derrida, J. (2002) 'The animal that therefore I am (more to follow)', *Critical Inquiry* 28: 369–418.

Derrida, J. (2003) 'And say the animal responded', in C. Wolfe (ed.) *Zoontologies: The Question of the Animal*, Minneapolis: University of Minnesota Press.

Desmond, A. (1998) *Huxley: From Devil's Disciple to Evolution's High Priest*, London: Penguin.

Desmond, A. and Moore, J. (1991) *Darwin*, London: Michael Joseph.

DETR (1999) *Towards an Urban Renaissance: Report of the Urban Task Force*, London: Department of Environment, Transport and the Regions.

Dickens, P. (1992) *Society and Nature: Towards a Green Social Theory*, Hemel Hempstead: Harvester Wheatsheaf.

Diprose, R. 2002 *Corporeal Generosity: On Giving with Nietzsche, Merleau–Ponty, and Levinas*, New York: State University of New York Press.

DLTR (2002) *Green Spaces, Better Places: Report from the Urban Green Spaces Task Force*, London: HMSO.

Donaldson, A. and Wood, D. 2004 'Surveilling strange materialities: categorisation in the evolving geographies of FMD biosecurity', *Environment and Planning D: Society and Space* 22: 373–91.

Driver, F. (2001) *Geography Militant: Cultures of Exploration and Empire*, Oxford: Blackwell.

Editorial (2006) 'Avian influenza goes global but don't blame the birds', *The Lancet: Infectious Diseases* 6(4): 185.

Evernden, N. (1992) *The Social Creation of Nature*, Baltimore, MD: Johns Hopkins University Press.

Fairhead, J. and Leach, M. (1998) *Reframing Deforestation: Global Analysis and Local Realities*, London: Routledge.

Fine, B. (2002) 'They f**k you up, those social capitalists', *Antipode* 34(4): 796–99.

Fitter, R. (1969) 'Foreword to the 1969 Edition', *Birds in London*, Newton Abbot: David and Charles.

Fitzsimmons, M. (1989) 'The matter of nature', *Antipode* 21(2): 106–20.

Foster, H. (2006) 'Go, modernity', *London Review of Books* 28(12): 11–12.

Foucault, M. (1970) *The Order of Things: An Archaeology of the Human Sciences*, London: Tavistock.

Foucault, M. (1973) *The Birth of the Clinic: An Archaeology of Medical Perception*, New York: Vintage.

Foucault, M. (1977) *Discipline and Punish: The Birth of the Prison*, Harmondsworth: Penguin.

Foucault, M. (1981) *The History of Sexuality,* Vol. 1: *An Introduction*, Harmondsworth: Penguin.

Franklin, A. (2002) *Nature and Social Theory*, London: Sage.

Fraser, M. (2004) 'Interdisciplinary, novel, and ethical relations', Paper given at *Life Sciences* conference, Queen Mary and Westfield College, University of London.

Fraser, M., Kember, S. and Lury, C. (2005) 'Inventive life: approaches to the new vitalism', *Theory, Culture & Society* 22(1): 1–14.

Fraser, N. (1986) 'Toward a discourse ethic of solidarity', *Praxis International* 5: 427–9.

Gandy, M. (2002) *Concrete and Clay: Reworking Nature in New York City*, Cambridge, MA: MIT Press.

Gatens, M. (1996) 'Through a Spinozist lens: ethology, difference, power', in P. Patton (ed.) *Deleuze: A Critical Reader*, Oxford: Blackwell.

Gertel, J. and Samir, S. (2000) 'Cairo: urban agriculture and "visions" for a modern city', in N. Bakker, M. Dubbeling, S. Gundel, U. Sabel-Koschella and H. de Zeeuw (eds) *Growing Cities, Growing Food: Urban Agriculture on the Policy Agenda. A Reader on Urban Agriculture*, Feldafing, Germany: German Foundation for International Developement (DSE).

Gibson-Graham, J.-K. (1996) *The End of Capitalism (as We Knew it): A Feminist Critique of Political Economy*, Oxford: Blackwell.

Giddens, A. (1998) *The Third Way: The Renewal of Social Democracy*, Cambridge: Polity Press.

Glacken, C. (1967) *Traces on the Rhodian Shore*, Berkeley, CA: University of California Press.

Goffman, E. (1971) *Frame Analysis: An Essay in the Organization of Experience*, Chicago: Northeastern University Press.

Goodwin, B. (1994) *How the Leopard Changed its Spots: The Evolution of Complexity*, London: Phoenix.

Graham, S. (2004) 'Cities as strategic sites: annihilation and urban geopolitics', in S. Graham (ed.) *Cities, War and Terrorism*, Oxford: Blackwell.

Grain (2006) 'The top-down global response to bird flu', *Against the Grain*, Barcelona.

Gray, J. (2002) *Straw Dogs*, London: Granta.

Gregory, D. (2004) *The Colonial Present*, Oxford: Blackwell.

Grundmann, R. (1991) 'The ecological challenge to Marxism', *New Left Review* 187: 103–20.

Hacking, I. (2000) 'Our fellow animals', *The New York Review of Books*.

Haraway, D. (1985) 'Manifesto for cyborgs: science, technology and socialist feminism in the 1980s', *Socialist Review* 80: 65–108.

Haraway, D. (1991a) *Simians, Cyborgs, and Women: The Reinvention of Nature*, London: Free Association Books.

Haraway, D. (1991b) 'Situated knowledges: the science question in feminism and the privilege of partial perspective', in D. Haraway (ed.) *Simians, Cyborgs and Women: The Reinvention of Nature*, London: Free Association Books.

Haraway, D. (1992) 'Otherworldly conversations; terrain topics; local terms', *Science as Culture* 3(1): 64–98.

Haraway, D. (1993) 'The promises of monsters: a regenerative politics for inappropriated others', in L. Grossberg, C. Nelson and P. Treichler (eds) *Cultural Studies*, London: Routledge.

Haraway, D. (1997) *Modest_Witness@Second Millennium: Femaleman_Meets-Oncomouse TM*, London: Routledge.

Haraway, D. (2003) *The Companion Species Manifesto: Dogs, People and Significant Otherness*, Chicago: Prickly Paradigm Press.

Harman, G. (2002) *Tool–being: Heidegger and the Metaphysics of Objects*, Chicago: Open Court.

Harrison, C. and Davies, G. (2002) 'Conserving biodiversity that matters: practitioners' perspectives on brownfield development and urban nature conservation in London', *Journal of Environmental Management* 65: 95–108.

Harvey, D. (1993) 'The nature of environment: dialectics of social and environmental change', in R. Miliband and L. Panitch (eds) *Socialist Register: Real Problems, False Solutions*, Oxford: Blackwell.

Harvey, D. (1996) *Justice, Nature and the Geography of Difference*, Oxford: Blackwell.

Hays, S. P. (1959) *Conservation and the Gospel of Efficiency*, Cambridge, MA: Harvard University Press.

Hearne, V. (1991) 'What's wrong with animal rights: horses, hounds and Jeffersonian happiness', *Harpers Magazine*.

Heidegger, M. (1978) 'The question concerning technology', in D. F. Krell (ed.) *Basic Writings*, London: Routledge and Kegan Paul.

Hetherington, K. (1997) *Badlands of Modernity*, London: Routledge.

Hetherington, K. and Lee, N. (2000) 'Social order and the blank figure', *Environment and Planning D: Society and Space* 18: 169–84.

Hinchliffe, S. (2000a) 'Living with risk: the unnatural geography of environmental crises', in S. Hinchliffe and K. Woodward (eds) *The Natural and the Social*, London/Milton Keynes: Routledge/Open University.

Hinchliffe, S. (2000b) 'Pigeons', in S. Pile and N. Thrift (eds) *City A–Z*, London: Routledge.

Hinchliffe, S. (2001) 'Indeterminacy in-decisions – science, policy and politics in the BSE crisis', *Transactions of the Institute of British Geographers* 26(2): 182–204.

Hinchliffe, S. (2003) 'Inhabiting – landscapes and natures', in K. Anderson, M. Domosh, S. Pile and N. Thrift (eds) *The Handbook of Cultural Geography*, London: Sage.

Hinchliffe, S. (2005) 'Nature/Culture', in D. Atkinson, P. Jackson, D. Sibley and N. Washbourne (eds) *Cultural Geography: A Critical Dictionary of Key Concepts*, London: I.B. Taurus.

Hinchliffe, S. (2007) 'Reconstituting Nature conservation: towards a careful political ecology', *Geoforum* 38.

Hinchliffe, S. and Blowers, A. (2003) 'Environmental responses: radioactive risks and uncertainty', in A. Blowers and S. Hinchliffe (eds) *Environmental Responses*, Chichester/Milton Keynes: John Wiley and Sons/Open University.

Hinchliffe, S., Kearnes, M., Degen, M. and Whatmore, S. (2005) 'Urban wild things: a cosmopolitical experiment', *Environment and Planning D: Society and Space* 23(5): 643–58.

Hinchliffe, S., Kearnes, M., Degen, M. and Whatmore, S. (2007) 'Ecologies and economics of action: Sustainability, Calculations and other things', *Environment and Planning A* 39(2): 260–82.

Hinchliffe, S. and Whatmore, S. (2006) 'Living cities: towards a politics of conviviality', *Science as Culture* 15(2): 123–38.

Hinchliffe, S. and Woodward, K. (2000) *The Natural and the Social: Uncertainty, Risk, Change*, London/Milton Keynes: Routledge/Open University.

Ingold, T. (1995) 'Building, dwelling, living: How animals and humans make themselves at home in the world,' in M. Strathern (ed.) *Shifting Contexts. Transformations in Anthropological Knowledge*, London: Routledge.

Ingold, T. (2000) *The Perception of the Environment: Essays in Livelihood, Dwelling and Skill*, London: Routledge.

Irwin, A. (1995) *Citizen Science: A Study of People, Expertise and Sustainable Development*, London: Routledge.

Jardine, N., Secord, J. and Spary, E. C. (eds) (1996) *Cultures of Natural History*, Cambridge: Cambridge University Press.

Jenkins, D. (ed.) (2005) *Catalogue: Foster and Partners*, London: Prestel.

Kant, I. (1948) *Groundwork of the Metaphysic of Morals*, London: Hutchinson University Library.

Keller, E. F. (1992) *A Feeling for the Organism: The Life and Work of Barbara McClintock*, San Francisco: W.H. Freeman.

Keyes, M. A. (1999a) 'The prion challenge to the 'Central Dogma' of molecular biology, 1965–1991. Part I: Prelude to prions', *Studies in the History and Philosophy of Biology and Biomedical Sciences* 30(1): 1–19.

Keyes, M. A. (1999b) 'The prion challenge to the 'Central Dogma' of molecular biology, 1965–1991. Part II: the problem with prions', *Studies in the History and Philosophy of Biology and Biomedical Sciences* 30(2): 181–218.

Lacey, R. (1994) *Mad Cow Disease: Dead End Host?*, Jersey: Cypsela.

Latour, B. (1987) *Science in Action: How to Follow Scientists and Engineers through Society*, Cambridge, MA: Harvard University Press.

Latour, B. (1988) *The Pasteurisation of France*, Cambridge, MA. : Harvard University Press.

Latour, B. (1993) *We Have Never Been Modern*, Hemel Hempstead: Harvester Wheatsheaf.

Latour, B. (1996) *Aramis, or the Love of Technology*, Cambridge, MA: Harvard University Press.

Latour, B. (1999) *Pandora's Hope: Essays on the Reality of Science Studies*, Cambridge, MA: Harvard University Press.

Latour, B. (2000) 'When things strike back – a possible contribution of science studies to the social sciences', *British Journal of Sociology* 51(1): 107–23.

Latour, B. (2002) 'Morality and technology: the end of the means', *Theory, Culture & Society* 19(5/6): 257–60.

Latour, B. (2003) *The War of the Worlds: What About Peace?*, Chicago: Prickly Paradigm Press.

Latour, B. (2004a) 'How to talk about the body? The normative dimension of science studies', *Body & Society* 10: 205–29.

Latour, B. (2004b) *Politics of Nature: How to Bring the Sciences into Democracy*, Cambridge, MA: Harvard University Press.

Latour, B. (2005) *Reassembling the Social: An Introduction to Actor-Network-Theory*, Oxford: Oxford University Press.

Latour, B. and Weibel, P. (eds) (2005) *Making Things Public. Atmospheres of Democracy*, Karlsruhe, Germany/Cambridge, MA: ZKM: Centre for Art and Media Karlsruhe/MIT Press.

Latour, B. and Woolgar, S. (1979) *Laboratory Life*, London: Sage.

Law, J. (1994) *Organizing Modernity*, Oxford: Blackwell.

Law, J. (2002) *Aircraft Stories: Decentering the Object in Technoscience*, Durham, NC: Duke University Press.

Law, J. (2004a) *After Method: Mess in Social Science Research*, London: Routledge.

Law, J. (2004b) 'Mattering, or how might STS contribute?' Centre for Science Studies, Lancaster University, Lancaster LA1 4YL, UK, http://www.comp.lancs.ac.uk/sociology/papers/law-mattering.pdf.

Law, J. (2006a) 'Disaster in agriculture: or foot and mouth mobilities', *Environment and Planning A* 38(2): 227–39.

Law, J. (2006b) 'Narratives and pinboards: on foot and mouth multiple' paper presented at Nature Politics Conference, University of Oslo, February.

Law, J. (2007) 'Pinboards and books: learning, materiality and juxtaposition', in D. Kritt and L. T. Winegar (eds) *Education and Technology: Critical Perspectives, Possible Futures*, Lanham, MD: Rowman and Littlefield.

Law, J. and Lynch, M. (1990) 'Lists, field guides, and the descriptive organization of seeing: birdwatching as an exemplary observational activity', in M. Lynch and S. Woolgar (eds) *Representation in Scientific Practice*, Cambridge, MA: MIT Press.

Law, J. and Mol, A. (eds) (2002) *Complexities: Social Studies of Knowledge Practices*, Durham, NC: Duke University Press.

Law, J. and Mol, A. (2007) 'Globalisation in practice: on the politics of boiling pigswill,' *Geoforum* 38.

Lee, N. and Brown, S. (1994) 'Otherness and the actor network: the undiscovered continent', *American Behavioural Scientist* 37: 772–90.

Lenoir, T. (1994) 'Was the last turn the right turn? The semiotic turn and A.J. Greimas', *Configurations* 1: 119–36.

Lewontin, R. (1993) *The Doctrine of DNA: Biology as Ideology*, Harmondsworth: Penguin.

Livingstone, D. (1992) *The Geographical Tradition*, Oxford: Blackwell.

Livingstone, D. (2003) *Putting Science in its Place: Geographies of Scientific Knowledge*, Chicago: University of Chicago Press.

Lyotard, J.-F. (1994) *Driftworks*, New York: Semiotext(e).

Macdonald, D. W., Mace, G. M. and Rushton, S. (2000) 'British mammals: is there a radical future?' in A. Entwistle and N. Dunstone (eds) *Priorities for the Conservation of Mammalian Diversity: Has the Panda Had its Day?*, Cambridge: Cambridge University Press.

Malthus, T. R. (1992) *Malthus: An Essay on the Principle of Population*, Cambridge: Cambridge University Press.

Massey, D. (1995) 'Thinking radical democracy spatially', *Environment and Planning D: Society and Space* 13: 283–88.

Massey, D. (1999) *Power Geometries and the Politics of Space-Time*, Heidelberg: Department of Geography, University of Heidelberg.

Massey, D. (2004) 'Geographies of responsbility', *Geografiska Annaler* 86(1): 5–18.

Massey, D. (2005) *For Space*, London: Routledge.

Massumi, B. (1996) 'The autonomy of affect', in P. Patton (ed.) *Deleuze: A Critical Reader*, Oxford: Blackwell.

Matless, D. (1998) *Landscape and Englishness*, London: Reaktion.

Maturana, H. and Varela, F. (1992) *The Tree of Knowledge: The Biological Roots of Human Understanding*, Boston: Shambala Press.

Mazis, G. A. (1999) 'Chaos theory and Merleau-Ponty's ontology: beyond the dead father's paralysis toward a dynamic and fragile materiality', in D. Olkowoski and J. Morley (eds) *Merleau–Ponty: Interiority and Exteriority, Psychic Life and the World*, Albany, NY: SUNY.

McKibben, B. (2003) *The End of Nature: Humanity, Climate Change and the Natural World*, 2nd edn, London: Bloomsbury.

McNay, L. (2000) *Gender and Agency*, Polity: Cambridge.

Merchant, C. (1990) *The Death of Nature: Women, Ecology and the Scientific Revolution*, New York: Harper and Row.

Merleau-Ponty, M. (1962) *Phenomenology of Perception*, London: Routledge and Kegan Paul.

Mitchell, T. (2002) *Rule of Experts: Egypt, Techno-politics, Modernity*, Berkeley, CA: University of California Press.

Mol, A. (1999) 'Ontological politics, a word and some questions', in J. Law and J. Hassard (eds) *Actor Network Theory and After*, Oxford and Keele: Blackwell/ Sociological Review.

Mol, A. (2002) *The Body Multiple: Ontology in Medical Practice*, Durham, NC: Duke University Press.

Mol, A. (forthcoming) 'The logic of care', unpublished MS.

Mol, A. and Law, J. (1994) 'Regions, networks and fluids: anaemia and social topology', *Social Studies of Science* 24: 641–71.

Mouffe, C. (1993) *The Return of the Political*, London: Verso.

Mouffe, C. (2000) *The Democratic Paradox*, London: Verso.

Munro, R. (1997) 'Ideas of difference: stability, social spaces and the labour of division', in K. Hetherington and R. Munro (eds) *Ideas of Difference*, Oxford and Keele: Blackwell and Sociological Review.

Murdoch, J. (2003) 'Geography's circle of concern', *Geoforum* 34: 287–89.

Murdoch, J. (2005) *Poststructural Geography: A Guide to Relational Space*, London: Sage.

ODPM (2005) *Planning Policy Statement 9: Biodiversity and Geological Conservation*, London: The Stationary Office.

ODPM and DEFRA (2005) *Government Circular: Biodiversity and Geological Conservation – Statutory Obligations and their impact within the Planning System*, London: The Stationary Office.

O'Riordan, T. (1976) *Environmentalism*, London: Pion.

Passmore, J. (1980) *Man's Responsibility for Nature*, 2nd edn, London: Duckworth.

Patton, P. (2003) 'Language, power and the training of horses', in C. Wolfe (ed.) *Zoontologies: The Question of the Animal*, Minneapolis: University of Minnesota Press.

Paulson, W. (2001) 'For a cosmopolitical philology: lessons from science studies', *SubStance* #96 30(3): 101–119.

Pennington, H. (2000) 'The English disease', *London Review of Books*.

Perrow, C. (1999) *Normal Accidents: Living with High Risk Technologies*, 2nd edn, Princeton, NJ: Princeton University Press.

Phillips, L., Bridgeman, J. and Ferguson-Smith, M. (2000) *The BSE Inquiry*, Vols I–XVI, London: HMSO.

Philo, C. (2005) 'Spacing lives and lively spaces: partial remarks on Sarah Whatmore's hybrid geographies', *Antipode* 37(4): 824–33.

Pile, S. and Thrift, N. (eds) (1995) *Mapping the Subject*, London: Routledge.

Pols, J. (2003) 'Enforcing patient rights or improving care? The interference of two modes of doing good in mental health care', *Sociology of Health and Illness* 25(4): 320–47.

Prusiner, S. B. and McKinley, M. P. (eds) (1987) *Prions. Novel Infectious Pathogens Causing Scrapie and Creuzfeldt-Jakob Disease*, San Diego: Academic Press.

Putnam, R. (2000) *Bowling Alone: The Collapse and Revival of American Community*, New York: Simon and Schuster.

Rajchman, J. (2000) *The Deleuze Connections*, Cambridge, MA: MIT Press.

Regan, T. (1984) *The Case for Animal Rights*, London: Routledge.

Rheinberger, H.-J. (1997) *Towards a History of Epistemic Things: Synthesizing Proteins in the Test Tube*, Stanford, CA: Stanford University Press.

Ridley, R. M. and Baker, H. F. (1998) *Fatal Protein: The Story of CJD, BSE and Other Prion Diseases*, Oxford: Oxford University Press.

Ripple, W. J. and Beschta, R. L. (2003) 'Wolf reintroduction, predation risk, and cottonwood recovery in Yellowstone National Park', *Forest Ecology and Management* 184: 299–313.

Rogers, R. and Gumuchdijan, P. (1997) *Cities for a Small Planet*, London: Faber and Faber.

Rose, S. (1998) *Lifelines: Biology, Freedom, Determinism*, London: Penguin.

Rose, S., Kamin, L. J. and Lewontin, R. (1984) *Not in Our Genes: Biology, Ideology and Human Nature*, Harmondsworth: Penguin.

Secord, J. (1981) 'Nature's fancy: Charles Darwin and the breeding of pigeons', *Isis* 72: 163–86.

Serres, M. (1995a) *Conversations on Science, Culture and Time*, Ann Arbor, MI: University of Michigan Press.

Serres, M. (1995b) *The Natural Contract*, Ann Arbor, MI: The University of Michigan Press.

Shapin, S. and Schaffer, S. (1985) *Leviathan and the Air Pump: Hobbes, Boyle and the Experimental Life*, Princeton, NJ: Princeton University Press.

Sheinin, D. (1994) 'Defying infection: Argentine foot and mouth policy, 1900–1930', *Canadian Journal of History* 29(3): 501–23.

Shusterman, R. (1997) *Practicing Philosophy: Pragmatism and the Philosophical Life*, New York: Routledge.

Singer, P. (1984) *Animal Liberation*, London: Jonathan Cape.

Slackman, M. (2006) 'Bird flu or not, Egyptians keep their ducks', *International Herald Tribune*, 31 May.

Slicer, D. (1991) 'Your daughter or your dog? A feminist assessment of the animal research issue', *Hypatia* 6(1): 108–24.

Soper, K. (1995) *What is Nature?*, Oxford: Blackwell.

Southwood, R. (1989) *Report of the Working Party on Bovine Spongiform Encephalopathy*, London: Ministry of Agriculture, Fisheries and Food.

Stengers, I. (1997) *Power and Invention: Situating Science*, Minneapolis: University of Minnesota Press.

Stengers, I. (2000) *The Invention of Modern Science*, Minneapolis: University of Minnesota Press.

Stengers, I. (2004) 'A constructivist reading of process and reality'. Paper given at Goldsmiths College, London available at www.goldsmiths.ac.uk/CSISP/papers/ stengers_constructivist_reading.pdf.

Strathern, M. (1991) *Partial Connections*, Savage, MD: Rowman and Littlefield.

Strathern, M. (1996) 'Cutting the network', *Journal of the Royal Anthropological Institute* 2: 517–535.

Swyngedouw, E. (2004) *Social Power and the Urbanization of Water*, Oxford: Oxford University Press.

Thacker, E. (2005a) 'Living dead networks', *FibreCulture* 4: http://journal. fibreculture.org/issue4/issue4_thacker.html.

Thacker, E. (2005b) 'Nomos, nosos and bios', *Culture Machine* 7: http://culturemachine. tees.ac.uk/frm_f1.htm.

Thompson, C. (2002) 'When elephants stand for competing philosophies of natures: Amboseli National Park, Kenya', in J. Law and A. Mol (eds) *Complexities: Social Studies of Knowledge Practices*, Durham, NC: Duke University Press.

Thrift, N. (2000) 'Still life in the nearly present time', *Body and Society* 6(3–4): 34–57.

Thrift, N. (2003) 'All nose', in K. Anderson, M. Domosh, S. Pile and N. Thrift (eds) *The Handbook of Cultural Geography*, London: Sage.

Thrift, N. (2005) 'From born to made: technology, biology and space', *Transactions of the Institute of British Geographers* 30: 463–76.

Tuan, Y. F. (1984) *Dominance and Affection: The Making of Pets*, New Haven, CT: Yale University Press.

Verran, H. (2001) *Science and an African Logic*, Chicago: Chicago University Press.

Verran, H. (2002) 'A postcolonial movement in science studies: alternative firing regimes of environmental scientists and Aboriginal landowners', *Social Studies of Science* 32 (5–6): 729-62.

Waterton, C. and Ellis, R. (2004) 'Environmental citizenship in the making: the participation of volunteer naturalists in the UK biological recording and biodiversity policy', *Science and Public Policy* 31(2): 95–105.

Watson, S. (1998) 'The new Bergsonism', *Radical Philosophy* 92: 1–23.

Weber, M. (1991) *From Max Weber: Essays in Sociology*, London: Routledge.

Western, D., Wright, M. and Strum, S. (eds) (1994) *Natural connections: Perspectives in Community-based Conservation*, Washington, DC: Island Press.

Whatmore, S. (1997) 'Dissecting the autonomous self: hybrid cartographies for a relational ethics', *Environment and Planning D: Society and Space* 15: 37–53.

Whatmore, S. (2002) *Hybrid Geographies: Natures, Culture, Spaces*, London: Sage.

Whatmore, S. (2003) 'Generating materials', in M. Pryke, G. Rose and S. Whatmore (eds) *Using Social Theory*, London: Sage.

Whatmore, S. (2004) 'Humanism's excess: some thoughts on the post-human/ist agenda', *Environment and Planning A* 36: 1360–3.

White, S. (1991) *Political Theory and Postmodernism*, Cambridge: Cambridge University Press.

WHO (2006a) *Questions and Answers on Avian Influenza*, Geneva: World Health Organization.

WHO (2006b) 'Avian Influenza Fact Sheet February (2006)', www.who.int/media centre/factsheets/avian_influenza/en/print.html.

Williams, G. (2001) 'Where the wild things are: an interview with Steve Baker', *Cabinet Magazine Online*: 1–7. Available at: http: //cabinetmagazine.org/issues/4/stevebaker. php.

Williams, P., Gaston, K. and Humphries, C. (1994) 'Do conservationists and molecular biologists value differences between organisms in the same way?' *Biodiversity Letters* 2(3): 67–78.

Wilson, A. (1992) *The Culture of Nature*, Oxford: Blackwell.

Wilson, E. A. (1996) 'On the nature of neurology', *Hysteric: Body/Medicine/Text* 2: 49–63.

Wolch, J. (1998) 'Zoopolis', in J. Wolch and J. Emel (eds) *Animal Geographies*, London: Verso.

Wolch, J. and Emel, J. (eds) (1998) *Animal Geographies*, London: Verso.

Wolfe, C. (2003a) 'In the shadow of Wittgenstein's lion: language, ethics, and the question of the animal', in C. Wolfe (ed.) *Zoontologies: The Question of the Animal*, Minneapolis: University of Minnesota Press.

Wolfe, C. (2003b) 'Introduction', in C. Wolfe (ed.) *Zoontologies: The Question of the Animal*, Minneapolis: University of Minnesota Press.

Wolfe, C. (ed.) (2003c) *Zoontologies: The Question of the Animal*, Minneapolis: University of Minnesota Press.

Woods, A. (2004a) 'The construction of an animal plague: foot and mouth disease in nineteenth–century Britain', *Social History of Medicine* 17(1): 23–39.

Woods, A. (2004b) 'Flames and fear on the farms': controlling foot and mouth disease in Britain, 1892–2001', *Historical Research* 77(198): 520–42.

Woods, A. (2004c) *A Manufactured Plague: The History of Foot and Mouth Disease in Britain*, London: Earthscan.

Worster, D. (1988) 'Doing environmental history', in D. Worster (ed.) *The Ends of the Earth: Perspectives on Modern Environmental History*, Cambridge: Cambridge University Press.

Young, R. M. (1985) *Darwin's Metaphor: Nature's place in Victorian Culture*, Cambridge: Cambridge University Press.

Index